Clinical Microbiology

Clinical Microbiology

P. W. Ross
TD, MD, MRCPath, MRCPE, MIBiol
Senior Lecturer, Department of Bacteriology,
University of Edinburgh Medical School;
Honorary Consultant,
Royal Infirmary, Edinburgh, UK

J. F. Peutherer
BSc, MD, MRCPath, MRCPE
Senior Lecturer, Department of Bacteriology,
University of Edinburgh Medical School;
Honorary Consultant,
Royal Infirmary, Edinburgh, UK

CHURCHILL LIVINGSTONE
EDINBURGH LONDON AND NEW YORK 1987

CHURCHILL LIVINGSTONE
Medical Division of Longman Group UK Limited

Distributed in the United States of America by
Churchill Livingstone Inc., 1560 Broadway,
New York, N.Y. 10036, and by associated companies,
branches and representatives throughout the world.

© Longman Group UK Limited 1987

All rights reserved. No part of this publication may
be reproduced, stored in a retrieval system, or
transmitted in any form or by any means, electronic,
mechanical, photocopying, recording or otherwise,
without the prior permission of the publishers
(Churchill Livingstone, Robert Stevenson House,
1-3 Baxter's Place, Leith Walk,
Edinburgh EH1 3AF).

First published 1987

ISBN 0-443-03333-1

British Library Cataloguing in Publication Data
Ross, Philip W.
  Clinical microbiology.
  1. Medical microbiology
  I. Title II. Peutherer, J. F.
  616'.01   QR46

Library of Congress Cataloging in Publication Data
Ross, Philip W.
  Clinical microbiology.

  Includes index.
  1. Medical microbiology. 2. Infection.
I. Peutherer, J. F. II. Title. [DNLM: 1. Infection—
microbiology. 2. Infection—prevention & control.
3. Microbiological Technics. 4. Microbiology.
WC 200 R825cb]
QR46.R764 1986   616.9   86-17158

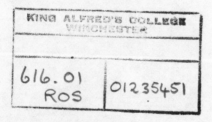

Produced by Longman Singapore Publishers (Pte) Ltd.
Printed in Singapore

Preface

This book aims to provide the essentials of medical microbiology for those involved in the diagnosis and management of infection, including medical students, hospital medical staff and general practitioners. To help clarify the approach to problems of infection the various body systems have been dealt with in separate chapters. This book focusses on the clinical illness or lesions rather than the causative organisms.

Clinical microbiology is concerned with the bacteria, viruses, chlamydiae, mycoplasmas, rickettsiae, fungi and protozoa that infect man, and this book deals with the following aspects: (a) how infections are spread amongst individuals and in the population at large; (b) how disease is produced; (c) the factors that determine the outcome of microbial challenge; (d) the control of infection by sterilisation, disinfection and immunisation; (e) antimicrobial chemotherapy.

Our thanks are due to Mr Ian Lennox, University of Edinburgh, for illustrative material, to Mrs Vivienne McGrath for typing the script and to Mrs Audrey Peutherer and Mrs Stella Ross for checking the text and reading the proofs.

Edinburgh, 1987

P. W. R.
J. F. P.

Contents

1. Sources and spread of infections	1
2. Microbial pathogenicity	13
3. The host response and prevention of infections	29
4. Diagnosis of infections	57
5. Treatment of infections	74
6. Infections of the respiratory tract	97
7. Infections of the cardiovascular, blood and lymphoreticular systems	118
8. Infections of the mouth	133
9. Infections of the gastrointestinal tract and liver	145
10. Infections of the urinary tract	168
11. Infections of the genital tract	177
12. Infections in pregnancy and the puerperium: congenital and neonatal infections	191
13. Infections of the central nervous system	199
14. Infections of the locomotor system	216
15. Infections of wounds, burns, skin and eye	221
16. Miscellaneous bacterial infections	238
Index	247

1

Sources and spread of infections

Definitions

The terms *source*, *reservoir* and *vehicle* of infections are frequently confused and misused. *Source* refers to the *habitat* or growth area in the human or animal. *Reservoir* and *vehicle* refer to objects that are contaminated or colonised by microorganisms. For example, fingers may be the vehicle or reservoir for staphylococci derived from the anterior nares; milk may be the reservoir of *Brucella* organisms the source of which can be an infected cow and water may act as a reservoir for cholera vibrios whose source is the human intestine.

Infections may be *endogenous* (autogenous) or *exogenous*. *Endogenous infections* are contracted from the host himself. Many areas of the body have a normal, commensal flora characteristic of the particular area. A most important entity, this flora has many functions including the provision of a barrier to infection in the individual. Normally the organisms that comprise the commensal flora do not cause infection in the host but there are exceptions to the rule.

For example, two species of the viridans streptococci, *Streptococcus sanguis* and *mitis*, may produce infection in previously damaged heart valves when they enter the bloodstream from the mouth following dental extraction. *Streptococcus faecalis* can cause *infective endocarditis* and the source of these organisms is usually the urinary tract and intestine of the host. Their presence in the blood and consequently the heart valves may be the result of diagnostic surgical procedures such as cystoscopy. Other examples of endogenous infection include osteomyelitis in which *Staphylococcus aureus* may have been derived from a septic lesion in the skin; infection of burns by beta-haemolytic streptococci from the patient's own upper respiratory tract and urinary tract infection by *Escherichia coli* derived from the intestine. There are three common features in endogenous infections. Firstly, the infections are produced some distance away from the normal habitat of the organisms; secondly, such infections are frequently a manifestation of lowered resistance, for example, the compromised patient, or tissue damage; and finally, problems of endogenous infections are generally confined to the patient, in that they do not generally constitute high cross-infection risks. Some viruses can cause a latent infection of the host and may be reactivated and excreted without clinical signs. However, infection of the skin and mucous membranes may occur after reactivation of herpes simplex virus and *recrudescent* lesions develop. This period of growth enhances virus concentration and can lead to excretion and transmission of infection.

Exogenous infections represent the greater proportion of infections and are derived from *Man, Animals* or the *Soil. Man* is an important source of exogenous infection, either when the patient is suffering from *clinical infection* or when the person is a *carrier* of infection. In whooping cough, smallpox and influenza where the carrier state is thought not to exist, or at least to be of minimal importance in the transmission of infection, the clinical or subclinical case is the source and such patients represent a major danger in the spread of these infections. The *incubation period* is the interval—months, weeks, days or hours—between exposure and the appearance of symptoms and signs of infection. Late in the incubation period, the organism has multiplied extensively in the organs and tissues; at this time the patient will be highly infectious. Many virus infections are inapparent or accompanied only by mild or non-specific symptoms. In poliomyelitis more than 90% of infections are unrecognised; however, patients in this state are infectious and are important in the spread of the disease.

Infections are commonly contracted from *carriers* and various types of these are described, such as *healthy, convalescent* and *chronic*. Healthy carriers are infected but have no signs of infection: if they excrete the organism, they are a particularly dangerous form of carrier. A good example is the healthy carrier of hepatitis B virus whose blood may carry virus over many years. Streptococcal sore throat, various types of pneumonia and meningococcal meningitis can be contracted thus, as can intestinal infections such as bacillary dysentery and the enteric fevers. *Contact carriers* are those who acquire the infecting organisms from a patient but who do not become clinically ill. *Convalescent carriers* are those who shed organisms for variable periods after clinical infection. There is no defined period for convalescent carriage. Studies have shown that after an attack of diphtheria organisms persist in the throat of the patient for a period of several weeks, whereas after an attack of bacillary dysentery the carrier state can remain for several months. Similarly, poliovirus can be excreted in the faeces for some weeks after clinical or sub-clinical infection. Administration of antimicrobial agents has frequently little or no effect in shortening the duration of carriage and in the case of intestinal infection may actually increase the duration. There is no clear-cut line of demarcation between the end of convalescent and the beginning of *chronic carriage*. In the case of typhoid fever, however, it is customary to describe carriage as chronic when more than one year has elapsed since the onset of the acute infection. With hepatitis B, a patient is a chronic carrier if the virus persists for more than 6 months. In general terms much has still to be learned about the dynamics of the carrier state.

The ability of an individual to spread infection in a community is frequently in inverse proportion to the seriousness of the infection. For example, carriers in the community are usually healthy, undetected and free to disseminate their organisms, whereas an infected, ill patient will contact his doctor who will either treat him at home or send him to hospital; he is *identified* as a case of infection. Alternatively, he may not seek medical help but he may at least stay off work or school; in the case of *serious infection* the patient will be *isolated* in hospital, and will have special nursing and other procedures; the risks of spread of infection are then minimal.

Animals are important sources of infection that may be transmitted to man; such infections are known as *Zoonoses*. Spread of these is usually from animal to animal. Man may be infected as an *end-host* and does not further spread the infection, as in rabies. In other instances the infected human may initiate man to man spread, as in pneumonic plague and yellow fever in the urban form. Examples of some diseases that can be acquired from animals can be seen in Table 1.1

Soil has also a role in the production of infections. This is particularly so in infections with *Clostridium tetani* and *Clostridium perfringens* in which soil is the reservoir rather than the source of infection; the source is generally the intestine of animals and the organisms gain access to the soil from animal faeces. Various fungal groups such as *Histoplasma* and *Microsporum* are also present in soil.

Table 1.1 Infections that may be acquired from animals, birds and fish

Human disease	Organism	Principal sources
Anthrax	*Bacillus anthracis*	Cattle, goats, sheep
Brucellosis	*Brucella abortus, suis, melitensis*	Cattle, goats, pigs, sheep
Bacterial food infections and intoxications	Certain campylobacters, clostridia, salmonellae *Staphylococcus aureus*	Cattle, pigs, poultry
Cryptococcosis	*Cryptococcus neoformans*	Cattle
Erysipeloid	*Erysipelothrix rhusiopathiae*	Fish, poultry, rodents
Lassa fever	Arenavirus	Mammals
Leishmaniasis	*Leishmania donovani*	Dogs
Leptospirosis	*Leptospira canicola* and *icterohaemorrhagiae*	Dogs, pigs, rodents
Listeriosis	*Listeria monocytogenes*	Birds, mammals
Lymphocyticchoriomeningitis	Arenavirus	Mice and hamsters
Marburg disease	Marburg virus	African green monkeys
Plague	*Yersinia pestis*	Rats
Psittacosis	Chlamydia	Birds (parrots and pigeons)
Q fever	*Coxiella burneti*	Cattle, goats, sheep
Rabies	Rhabdovirus	Bats, dogs and carnivores (foxes)
Ratbite fever	*Spirillum minus* *Streptobacillus moniliformis*	Rats
Ringworm	*Microsporum canis* *Trichophyton verrucosum*	Cats, dogs Cattle
Schistosomiasis	*Schistosoma japonicum*	Cattle, dogs, rodents
Tapeworm infection	*Taenia* spp	Cattle, pigs
Toxocariasis	*Toxocara canis*	Dogs
Toxoplasmosis	*Toxoplasma gondii*	Birds, mammals
Trypanosomiasis	*Trypanosoma* spp	Wild mammals
Tuberculosis	*Mycobacterium bovis*	Cattle
Yellow fever	Togavirus	Monkeys

OUTBREAKS OF INFECTIONS

Infections in humans may be *endemic, epidemic* or *pandemic*.

Endemic infections are those normally present in a small proportion of the community and in the United Kingdom bacillary dysentery is a good example of this. The endemic nature of infectious diseases depends on general factors such as social and economic conditions, on environmental factors such as population density and movement, standards of hygiene and sanitation and on general herd immunity. Generally herd immunity can be kept at a satisfactory level by immunisation procedures, as for example with measles, but clinical and subclinical infections are also important factors. *Exotic* infections result from the introduction of the disease from other countries. Cholera, lassa fever and malaria in the United Kingdom are such examples.

Epidemic infections result from any significant increase in the usually low incidence of endemic infections in a community. Many present-day epidemics particularly of virus infections such as influenza, measles and poliomyelitis are periodic. This periodicity is partly accounted for by an increase in the number of susceptible persons in the community as may occur when vaccination rates fall. Epidemics may also be caused by a breakdown in hygiene, a change in the virulence of existing organisms, or the introduction to the community of a new organism as in Fiji in 1875 when the island's first epidemic of measles caused 20 000 deaths.

Pandemic infections are epidemics that have reached massive proportions. They are national or international rather than community problems. Such outbreaks spread quickly from country to country aided by the rapidity of modern travel. The pandemics of Asian and Hong Kong influenza in 1957 and 1968 are recent examples.

When reference is made to infections in animal populations the terms *enzootic* rather than endemic, and *epizootic* rather then epidemic are used.

ROUTES OF SPREAD OF INFECTIONS

There are 5 main *routes* by which a host may become infected: the respiratory tract; the alimentary tract; the genital tract; the skin and muscosae; and the placenta; within these routes microorganisms can spread via the blood, lymphatics, nerves or by direct extension; viruses and bacterial toxins may spread via the peripheral nerves.

Spread of infections by the respiratory tract

Organisms that produce specific respiratory tract infections such as the common cold, influenza and pneumonia enter the body by this route, as well as organisms that cause generalised systemic infections such as chickenpox and measles. Three main methods are recognised in the spread of respiratory tract infections: *contact; direct airborne*; and *indirect airborne*.

Contact

This can be direct or indirect. In *direct* contact such as kissing, infected droplets are passed from person to person; this is an important means of spread of pathogens that are poorly viable outside the body, for example, glandular fever. In whooping cough the secondary attack rate in families may be as high as 90%, whereas in less intimate situations such as in hospital wards the secondary attack is much lower. In *indirect* contact, organisms are transferred from inanimate reservoirs of infection (fomites) such as cups and eating utensils. Although the division is made into direct and indirect contact to distinguish between direct droplet spread from an infected person and the spread from an inanimate reservoir of infection, it is frequently impossible to tell which has been the means of transfer of organisms to the susceptible host.

Direct airborne spread

Droplets are expelled from the mouth in talking and coughing and to a much greater extent in sneezing, and are conveniently classified as *large* and *small droplets*. They are derived mostly from the saliva and only a proportion will be infected. *Large droplets* have a diameter of over 100 μm and having been expelled from the mouth fall quickly to the ground and on to surfaces and clothing. Most fall within 6 feet of the person. *Small droplets*, less than 100 μm in diameter, constitute the main part of the droplet mass. They evaporate rapidly and become minute secretions with a diameter of 5–10 μm, called *droplet nuclei*, that may or may not contain organisms. These particles being light can remain suspended in air currents for many hours and are probably the means of spread of many virus infection; however, to transmit in this way the organism must be resistant to drying (Fig. 1.1).

Although droplet nuclei are considered important in the transmission of certain virus infections the role of the larger particles in producing direct airborne infection is less clear. Their importance may lie in their ability to contaminate the environment.

SOURCES AND SPREAD OF INFECTIONS 7

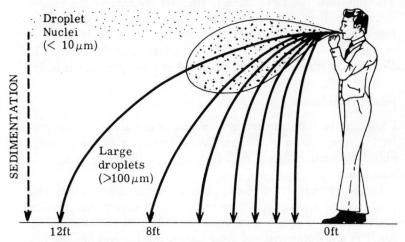

Fig. 1.1 Spread of droplets and droplet nuclei from the respiratory tract

Indirect airborne spread

When large respiratory droplets fall from the air they land on exposed surfaces, clothing and floors and become dried particles of secretions in the dust. Some of these will be infectious secretions and if protected from sunlight organisms will remain viable for weeks or months. *Mycobacterium tuberculosis* and *Streptococcus pyogenes* are good examples of this and these organisms can be released into the air by activities such as sweeping, dusting and bedmaking. Also, when a handkerchief is shaken dried particles of potentially infectious secretions are released into the air. Vesicular rashes, due to viruses, break down and shed virus into the air and the environment. Material shed from the lesions of chickenpox, shingles and smallpox can be inhaled to establish infection.

Whereas there may be one main mode of spread for most infections other routes may be responsible from time to time. For example, transmission of beta-haemolytic streptococci that cause sore throat may be by direct contact, indirect contact from fingers or clothing, or by dust. Similarly the common cold viruses may be spread via the fingers: the route is thus nose to finger, then to finger and nose of the recipient. Very occasionally respiratory infection, for example streptococcal sore throat, can be spread by consuming contaminated food and some cases of diphtheria have been attributed to drinking infected milk.

Spread of infections by the alimentary tract

The intestinal diseases, cholera, bacillary dysentery, the enteric fevers and bovine tuberculosis, are contracted when organisms are ingested, but the alimentary tract is also the route of entry of organisms such as enteroviruses, infectious hepatitis virus and *Brucella* spp whose effects are produced in other parts of the body.

Food contamination

Food may be contaminated at source or at any stage of its manufacture, preparation or storage. Contamination at source is exemplified by the infection of milk from a cow suffering from *Brucella* infection.

In the preparation of food, such as cream cakes, trifles and processed meat, organisms from many diverse sources can produce contamination. *Staphylococcus aureus* from the nose and skin of food handlers and *Salmonella* spp from hands, rodents and flies are but two examples. The symptomless carrier of salmonellae who is also a food handler is a dangerous source of infection. Milk is another potential danger particularly if it is not pasteurised. In the United Kingdom as much as 10–15% is still unpasteurised in which case the individual runs the risk of infection by organisms such as *Brucella abortus, Streptococcus pyogenes*, salmonellae, staphylococci and *Coxiella burneti*. Improper storage of the food will allow growth of the organism thus increasing the dose consumed. Viruses can only be transferred passively via food.

Water contamination

Infections may result from drinking water contaminated by urine or faeces, water being the reservoir and not the source. In most areas of Europe piped-water supplies have largely abolished waterborne infections though there is a risk in areas not yet within reach of a public supply. Classical Asiatic cholera and typhoid fever are associated with transmission by water. The Aberdeen typhoid outbreak in 1964, although disseminated by contaminated canned beef, was really an example of a water-borne infection because water used in cooling the cans after their sterilisation was polluted with sewage and several cans were improperly sealed, resulting in the contamination of the meat. Sewage contamination of shellfish is an important means of transmission of hepatitis A and some gastroenteritis viruses.

Spread of infections by hands

This type of spread is the result of inadequate personal hygiene and the source is usually a carrier who contaminates his hands at defaecation; even a wad of several sheets of tiolet paper, especially the absorbent varieties, may not be a barrier to organisms gaining access to the hands. These carriers contaminate toilet chains, wash basins, towels and door handles with their hands and infection is transmitted by another person putting his fingers to his mouth after handling contaminated articles. This is the means of spread of bacillary dysentery and poliomyelitis. Such spread is enhanced by the separation of the toilet and the wash hand basin.

Spread of infections through the skin and mucosae

Simple contact

Some infections can be contracted through the skin and mucosae, for example herpes simplex and hepatitis B. The common wart, impetigo and ringworm can be similarly spread either by direct contact with the infected person or by contact with his clothing, or other fomites contaminated by the virus as in gymnasia and swimming pools.

Contamination of wounds

A wound is taken as any breach of the surface of the skin and mucosae. Wounds resulting from trauma may be colonised by *Clostridium tetani*, particularly if contaminated by earth or dust. Abrasions and pricks by rose thorns can be similarly infected.

Sepsis of surgical wounds is one of the major infection problems of present-day hospitals and the causes of these are legion. A surgical wound is susceptible to infection from the moment of incision until it is completely healed and it is therefore exposed both in the operating theatre and in the ward. A vast literature has accumulated on wound sepsis in hospitals which highlights the roles of cases and carriers, operative and other surgical procedures, dressing techniques and environment contamination in its production (Ch. 15).

Injection

Injection of organisms can result from *medical* and *'recreational' injections* and *animal* or *insect bites*.

Medical injection and other invasive procedures, including blood transfusion, immunisation, prophylactic and therapeutic procedures, are seldom involved in the production of infection, although there is a well-known and very real risk of transmitting serum hepatitis by re-using needles and instruments contaminated with human blood. The recently identified virus associated with the acquired immune deficiency syndrome (AIDS) can also be transmitted by injection of blood and some blood products. Outwith the medical field severe septicaemia can be caused in drug addicts who make up their preparation for injection with unsterile water. A wide variety of organisms, including those of faecal origin, has been isolated from these. Addicts who share needles and syringes are also at great risk of acquiring hepatitis B from carriers.

Animal bites. Rabies and rat-bite fever are examples of infections introduced by animal bites, but many other organisms whose sources are animals can set up infection if the bite introduces them in sufficiently high numbers and if suitable conditions for multiplication are present.

Insect bites. Arthropods are important vectors of disease that in nature affect animals primarily and man only incidentally. All of the togaviruses are spread in this way and the virus inoculum is enhanced by multiplication of the virus in the insect vector. This is also true of malaria and trypanosomiasis. With some infections, for example yellow fever, man-to-man transmission can occur via the mosquito and can reach epidemic proportions.

Table 1.2 indicates some vectors involved in the spread of disease to man.

Table 1.2 Arthropod vectors of disease

Fleas	—Bubonic plague
Lice	—Epidemic typhus —European relapsing fever —Trench fever
Mites	—Rickettsial pox —Scrub typhus
Mosquitoes	—Many infections including dengue, malaria and yellow fever
Ticks	—African relapsing fever
Tsetse flies	—Trypanosomiasis

Spread of infections through the genital tract

Many infections spread by this route are sexually transmitted but those such as endometritis or cervicitis may be examples of endogenous infections.

Spread of infections across the placenta

An uncommon route of spread, the only bacterial infection in which this has been clearly established is syphilis in which the infected mother transmits *Treponema pallidum*. Transplacental spread is more common in certain viral diseases, notably rubella and cytomegalovirus infection and with the protozoon *Toxoplasma gondii*.

THE CONTROL OF SPREAD OF INFECTIONS

Although particular and specific measures can be applied to each infection, there are nevertheless general principles for the control and eradication of infections.

Isolation or eradication of source

Isolation measures are useful in both humans and animals with clinical or suspected infections. In many cases this approach is impossible as infection is usually unrecognised. In infections such as hepatitis B and AIDS isolation could only be achieved by a major change in the social and sexual behaviour of the carrier. Eradication of many bacterial infections is possible by the use of antimicrobial drugs. Removal of organisms by sterilisation procedures is routine in medical practice but many biological products cannot be sterilised.

Blocking routes of spread

To control airborne infection is difficult, although adequate ventilation of rooms and avoidance of overcrowding at times when respiratory disease are prevalent may have some value. Formerly floors of hospitals and barracks were oiled to minimise infection from dust, but whether this was beneficial is open to doubt. Positive pressure ventilation in operating theatres, isolation units and treatment rooms, has proved to be of value.

In the control of gastrointestinal diseases many measures can be taken, including maintaining purity of water supplies; ensuring pasteurisation of milk; maintaining scrupulous hygiene in the preparation, handling and storage of food, together with bacteriological monitoring of food handlers and effective disposal of excreta.

Protection of susceptible individuals

Immunisation procedures have proved highly successful in the control of infections such as smallpox, poliomyelitis, diphtheria and tetanus. Eradication of some diseases is possible, for example smallpox, if man is the only host and there is an effective vaccine. The administration of antimicrobial drugs may occasionally be useful in preventing certain infections in particularly susceptible individuals, for example the use of phenoxymethyl penicillin in the prevention of second attacks of rheumatic fever in children and adolescents. Great caution must be exercised however in administering antimicrobial drugs in the absence of firm bacteriological indications.

Finally, the resistance of patients to infection can be enhanced by general measures, for example proper nutrition and the control of diabetes.

FURTHER READING

Burnet F M 1962 Natural history of infectious diseases. Cambridge University Press, Cambridge
Duguid J P, Marimion B P, Swain R H A 1978 Medical microbiology. Churchill Livingstone, Edinburgh
Dubos R J, Hirsch J G 1965 Bacterial and mycotic infections of man, 4th edn. Pitman, London
Passmore R, Robson J S 1973 A companion to medical studies 2, revised reprint. Blackwell Scientific Publications, Oxford
Wilson G S, Miles A A 1975 Topley and Wilson's Principles of bacteriology and immunity, 6th edn. Arnold, London

2

Microbial pathogenicity

Pathogens

Pathogens are microorganisms that can cause disease in man and animals. *True* pathogens are those which by virtue of their properties frequently produce infection when in the host, although this is not inevitable since the host may be immune to the infecting organism; there are many other factors involved in the host-pathogen relationship. *Opportunist* pathogens are those which do not cause infection in the normal course of events but which may infect the *compromised host*—the person who suffers from hormonal, metabolic or malignant disease and the patient who is receiving either cytotoxic drugs or steroid hormones. Those undergoing immunosuppressive therapy are also very susceptible to infection with a wide variety of organisms because of derangement of the normal defence mechanisms of the body.

Virulence

This is a measure of the pathogenicity of the organism. All pathogens are endowed with some degree of virulence which depends on many factors including *structural components*, such as capsules, and *diffusible products*. Successful attempts have been made to measure the virulence of organisms in animals and the minimum lethal dose (MLD) and median lethal dose (LD_{50}) can be calculated. The MLD is ascertained by injecting serial dilutions of the organisms and finding the smallest dose required to kill the animal. As with humans however the resistance of animals can vary to a considerable extent and the LD_{50}, or the dose required to kill approximately 50% of the group of animals, may be a more useful measure of virulence. Death may not necessarily be the best criterion. The LD_{50} measured in an animal host is not a guide to the virulence in man. There are many examples of viruses that can only infect man and cannot be transmitted to other species. With some viruses this is due to the lack of receptors on the cell membrane.

In general, there are many factors of importance in determining the outcome of virus challenge. Some are dependent on the organism, others on the resistance of the host. Despite their small number of genes, it is not possible to ascribe virulence of a virus to one gene product as several are needed to allow viral infection and entry to target organs, i.e. the expression of virulence.

TRANSMISSIBILITY OF INFECTION

There are several factors involved in the transmission of infection from the original to the new host; for example, in the transmission of whooping cough close proximity to the original case is necessary. The *size of the infecting dose* is also important because successful transmission depends on having a sufficient number of organisms present to enter the body, overcome the host resistance and become established in the tissues. The dose of organisms required to initiate infection can be altered by certain factors such as the presence of foreign bodies in tissues and experiments have shown that the threshold number of organisms required to initiate a staphylococcal skin lesion can be reduced considerably from the several million required in normal circumstances if foreign bodies such as skin sutures are present.

Survival of organisms outside the body

This form of survival is important in the natural history of several infections. Spore-forming pathogens such as *Bacillus anthracis* and *Clostridium tetani* are notable for their survival outside the body, but more delicate organisms and strict parasites can survive for only limited periods under normal conditions. Most viruses are inactivated rapidly outside the body: some, for example, the Epstein–Barr virus (EBV) require direct transfer to infect, while poxviruses are much more stable and can survive in dried crusts for some time.

Survival of organisms in immune individuals

This is an important factor particularly in infections such as typhoid in which the bacilli can localise in the biliary tract, and related to this is the situation where persons become carriers without ever having had the clinical illness.

Survival of organisms in vectors or intermediate hosts

Bubonic plague is spread by rat fleas and a critical factor in the chain of events leading to infection of new hosts is the ability of the plague bacilli to infect the fleas.

EVENTS AFTER THE BACTERIAL ATTACK

Events depend largely upon the nature of the host-parasite association but the effects are essentially *localised* or *generalised*.

Localised infection

Many infections remain localised and only become generalised if the organism becomes more virulent, if the host resistance is decreased or if the bacteria gain access to another susceptible part of the body. The staphylococcal boil or furuncle typifies the localised infection, although many other bacteria can produce local lesions. Gonococci affect the urethra and conjunctiva and shigella organisms the intestinal mucosa. Some viruses stay at the point of entry where local multiplication produces the lesion, for example warts and molluscum contagiosum viruses in the skin.

Localised infection with diffusion of toxin

Diphtheria and streptococcal infections of the throat produce frank local infection with concomitant symptoms and signs as well as systemic involvement due to release of toxin. Other such infections include tetanus in which there may be only minimal local signs of bacterial growth.

Local spread and spread to adjacent tissues

The outcome of this depends on the result of the host-parasite encounter. Organisms that spread in the skin and subcutaneous tissues are frequently those that produce factors that facilitate their spread, for example streptokinase in the case of streptococcal infections. The mucus layer and secretions on mucosal surfaces, such as in the intestine or urethra, are important in spreading infection and establishing new foci in different areas.

Direct spread is not limited to skin and mucosal surfaces and there are many examples of direct spread within organs and tissues. In infection of the middle ear, for example, spread can occur to neighbouring structures such as the meninges and any focus of infection in the pelvis such as salpingitis can spread to involve the peritoneum. Similarly, pleurisy can result from an extension of lobar pneumonia. The mucus layer and secretions on mucosal surfaces such as in the intestinal and respiratory tracts are important in spreading viruses within these systems. Thus the symptoms of influenza and the common cold and viral gastro-enteritis appear when the causal agents have spread extensively within the systems from the initial focus of infection.

Generalised infection

Some organisms are highly invasive when they enter the body and produce a generalised infection. Some are classically acute as in the case of enteric fevers and others such as brucellosis may be subacute although acute episodes of the latter can occur. Spread can be direct, by the cerebrospinal fluid, by the blood or by lymphatics; certain organisms have a predilection for a particular organ or type of tissue, for example rabies virus and the central nervous system.

Spread by the cerebrospinal fluid

Bacteria in the blood can enter the cerebrospinal fluid (CSF) through the choroid plexus of the ventricles. Once within the subarachnoid space organisms spread very quickly because the CSF is devoid of any effective antibacterial properties. *Neisseria meningitidis, Haemophilus influenzae,* pneumococci and *Mycobacterium tuberculosis* enter the CSF from the bloodstream although bacteria may also gain access directly, as in the case of trauma to the skull or congenital deformities of the spinal cord. Similarly viruses can enter the CSF from the choroid plexus and meninges or from infected tissue.

Spread by the blood

The blood is an efficient means of disseminating organisms borne free in the plasma or within the white blood cells. Those carried free include staphylococci, streptococci and coliform organisms; such organisms are vulnerable to attack from the various defence mechanisms of the host. Bacteria carried by the white cells, particularly monocytes and lymphocytes, include *Brucella* spp and *Mycobacterium tuberculosis* and these are largely protected from the host defence factors. *Bacteriaemia* is simply the presence of bacteria in the blood and is a common occurrence. It can be produced merely by brushing teeth, or chewing, or by diagnostic procedures such as sigmoidoscopy and urinary tract catheterisation. It is generally a transient phenomenon as the bacteria are not actively multiplying and are quickly removed by the action of the host defences. Conversely, when large numbers of bacteria enter the bloodstream and actively multiply a *septicaemia* is produced. This is quite different from a simple bacteriaemia because the patient is clinically ill, often with focal infection in some organ, and the infection will progress unless controlled by antimicrobial drugs, with or without surgical intervention.

A viraemic phase occurs in all generalised virus infections as released virus or virus-infected monocytes or lymphocytes spread via lymphatics to the blood and thence to distant sites. Poliovirus spreads from the gut to the CNS via a viraemia. In contrast, measles virus circulates in monocytes and then localises in the skin to produce the characteristic rash. Viruses that circulate within the blood cells are protected from neutralisation by antibody. Microorganisms which are injected directly into the tissues, for example the malaria parasite, togaviruses and hepatitis B may stay within the lymphoreticular system or may invade target organs such as the liver and CNS.

Localisation of viruses

It is difficult to explain the different clinical patterns produced by viruses, for example why rash is the prominent feature of some infections whereas others affect the liver, or the meninges.

A number of explanations have been offered for the localisation of viruses. It is likely that most viruses that infect locally within the respiratory tract grow best at about 32°C and less effectively at 37°C. The same explanation is offered for the apparent localisation of smallpox and other viruses in the skin. For other sites it is suggested that the presence of receptor groups in the cell membrane is important, as is the association of poliovirus and the anterior horn cells of the spinal cord. The localisation may be more apparent than real as virus may be present at many sites. If the defensive responses of the host are inhibited or absent then virus may infect and damage many organs.

Spread by lymphatics

Bacteria can invade the lymphatic system in almost any part of the body but in most cases the lymph glands draining the various areas filter off the organisms and prevent further spread centrally. Some particularly virulent bacteria may not be filtered off and killed by the lymph glands and may spread to other areas of the body; indeed, some organisms may multiply actively within the gland. Organisms that spread via the lymphatic system include *Mycobacterium tuberculosis*, *Yersinia pestis*, *Brucella* spp and *Salmonella typhi*.

Spread along nerves

This is an extreme example of spread within a system. Some viruses, such as rabies, enter small nerve fibres at the site of injury and then ascend to the cord and brain. A similar sequence occurs with herpes simplex and varicella zoster viruses, except that virus ascends sensory nerve fibres only as far as the sensory root ganglia.

TOXIGENICITY

Bacterial toxins can be broadly divided into those secreted by bacteria during active growth, the *exotoxins*, and those associated

Table 2.1 A comparison of exotoxins and endotoxins

Exotoxins	Endotoxins
Produced by Gram-positive bacteria	Produced by Gram-negative bacteria
Protein	Lipopolysaccharide
Heat-labile	Heat-stable
Liberated from cytoplasm of multiplying bacteria	Liberated from cell wall of dead or disintegrating bacteria
Converted into toxoid by H.CHO	Cannot be toxoided
Highly specific for certain tissues	Non-specific
High potency	Low potency
Strongly antigenic	Poorly- or non-antigenic.
Effectively neutralised by antitoxin	Not effectively neutralised by antitoxin

with the structure of the organism such as the cell wall which are released after death of the cell, the *endotoxins*. Properties of toxins are shown in Table 2.1.
   Table 2.1 represents the classical view of differentiation of the toxins but is not strictly accurate. For example, protein toxins of some bacteria may be produced intracellularly as well as extracellularly during active growth and thus a new classification has been suggested, as shown in Table 2.2.

Table 2.2 Classification of bacterial toxins. (From Raynaud & Alouf 1970)

Group 1: Intracytoplasmic protein toxins of Gram-negative bacteria.

Group 2: Endotoxins of cell walls of Gram-negative bacteria.

Group 3: True protein exotoxins.

Group 4: Protein toxins with both intracellular and extracellular location during logarithmic phase.

Exotoxins

These are produced by several groups of Gram-positive bacteria such as *Clostridium botulinum, Clostridium perfringens* and *Clostridium tetani, Corynebacterium diphtheriae, Streptococcus pyogenes* and *Staphylococcus aureus* and Gram-negative bacteria such as *Shigella dysenteriae* and *Vibrio cholerae*. Many have affinities for particular tissues; for example, tetanus toxin affects the anterior columns of the spinal cord, diphtheria toxin the heart and peripheral nerve endings and staphylococcal toxin the vomiting centre in the central nervous system. Effects of toxins are *local* or *systemic*. Toxins acting locally include those produced by bacteria that gain access to the intestine by ingestion and exert their effect on small blood vessels and the intestinal mucosa, producing small ulcers as the result of toxic destruction of the epithelial cells. *Vibrio cholerae* releases a toxin that attaches to specific receptors on the epithelial cells, causing a rise in the level of cyclic adenosine monophosphate, thereby stimulating excessive secretion of water and electrolytes from the undamaged cells of the intestinal mucosa.
   Exotoxins that exert their effects at a distance are produced by *Corynebacterium diphtheriae, Clostridium tetani* and *Streptococcus pyogenes*. There is some controversy as to whether the typical skin rash of scarlet fever is due to a primary effect of the erythrogenic toxins or whether it is a manifestation of a hypersensitivity reaction to the toxins.

Toxoid

This is toxin that has been treated with certain chemicals such as formalin resulting in a loss of toxicity without altering its antigenicity. A toxoid is a good antigen and produces antitoxic antibody in man.

Endotoxins

These are complex polysaccharide-protein-phospholipid molecules associated with the cell walls of Gram-negative bacteria including *Escherichia coli*, *Proteus* spp, salmonellae, shigellae and *Bacteroides* spp and they produce many toxic effects in the body, particularly in the vascular system, causing a drastic fall in blood pressure. They are also very powerful pyrogens.

Endotoxic shock

Shock is often associated with a Gram-negative septicaemia related to the release of endotoxin from the bacteria in the bloodstream. Whether Gram-negative or bacteriogenic shock is synonymous with endotoxic shock is doubtful as some patients suffer Gram-negative shock with no evidence of circulating endotoxin. In addition, a syndrome similar to endotoxic shock can be caused by organisms that do not possess endotoxin. There may be some overlap in the causes of bacteriogenic and endotoxic shock although separate mechanisms may operate for each; the results are clinically similar, and many clinicians use the term 'septic shock'.

Patients with bacteriogenic or endotoxic shock are frequently those who have undergone a diagnostic procedure such as instrumentation of the urinary tract or they may have an active infection in the form of urinary tract infection or cholecystitis. Non-septic patients such as those who have suffered burns or trauma may be similarly affected.

Endotoxins may be detected in the blood of patients after instrumentation of the urinary tract with no subsequent deleterious effects and clinical manifestations may therefore be dose-related; the dramatic effects on the vascular system may occur only after large amounts of endotoxin enter the bloodstream. In addition to the effects on the vascular system by way of peripheral vascular pooling, drastic fall in blood pressure, collapse and possible death, endotoxin can be a cause of intravascular thrombosis. The condition is known as *disseminated intravascular coagulation* and is the result of clumping of the red blood cells consequent to an increase in capillary permeability and loss of fluid from the vascular system, together with activation of the blood clotting mechanism. This condition can also occur during some viral infections: in these, the trigger is believed to be the production of viral antigen-antibody complexes in the blood.

There are no known toxins of viruses, although some viral components may have local and systemic effects. However, the components released from damaged cells can induce the inflammatory response and interferons released from infected cells have local and systemic effects.

Many viruses damages cells directly. In several cases, for example polio, there is inhibition of host protein synthesis and hence a run-down of all cell functions. Eventually, leakage of lysosomal enzymes into the cytoplasm causes rupture of the cytoplasmic membrane. The histological changes associated with virus growth include cell rounding and fusion, although more specific changes can be seen, for example the presence of intranuclear inclusions (Cowdry type A) in cells infected with various herpesviruses and the cytoplasmic inclusions (Negri bodies) of rabies. However it is important to remember that some viruses do not kill the host cell and virus and cell can co-exist despite the production of progeny virions. Other viruses, the oncogenic viruses, alter the behaviour of the host cell and transform it. In addition to these direct effects, the presence of viral antigens on cell surfaces can lead to lysis of the cell by a variety of immunological mechanisms (see Ch. 3).

AGGRESSINS

Whether they produce infection by toxigenicity, invasiveness or a combination of both, bacteria possess structural components and other factors in the cytoplasm that contribute to the production of infection. These substances such as collagenases are not pathogenic in their own right but may nevertheless promote the pathogenicity of the organism. Factors that promote the initiation and spread of infection are often termed '*aggressins*', whether or not they themselves possess any toxic activity. Examples of aggressins are as follows:

Surface components

Prominent amongst these are bacterial *capsules*. Pneumococci possess polysaccharide capsules that help resist phagocytosis and the action of bactericidal substances in the body fluids; on the other hand, antibodies to these capsules assist phagocytosis and protect against infection. Other capsulate organisms are *Klebsiella pneumoniae* and *Haemophilus influenzae*.

Other surface components include the non-antigenic hyaluronic acid capsule of *Streptococcus pyogenes* but this is less effective in protection against phagocytosis than another surface component, the M-protein. Certain Gram-negative bacilli, particularly the enterobacteriaciae, form a layer of acidic polysaccharide outside their cell walls, which although not sufficiently substantial to be termed a capsule has nevertheless certain antiphagocytic properties.

Diffusible products

Many diffusible products in addition to the exotoxins can act as aggressins.

Hyaluronidase

Sometimes known as the *spreading factor* this is produced by *Staphylococcus aureus*, beta-haemolytic streptococci and *Clostridium perfringens*. It acts by breaking down the intercellular hyaluronic acid that binds cells together, allows spread of bacteria through the tissues and is thought to be important in the initiation of infection. There are anomalies in that *Staphylococcus aureus* produces large amounts but invasiveness and spread are not the hallmarks of typical staphylococcal skin infection and in infection by *Clostridium perfringens* antibodies to hyaluronidase neither prevent initiation of infection nor check its progress.

Coagulase

This substance is produced by *Staphylococcus aureus*. It is an enzyme that produces a fibrin envelope round the organisms conferring partial protection from phagocytosis. As yet no role has been found for coagulase in the production of infection and its usefulness may be more as a marker in the laboratory identification of pathogenic staphylococci.

Fibrinolysin

Some pathogenic staphylococci, groups A, C and G beta-haemolytic streptococci and *Bacteroides* spp produce this substance. A synonym for staphylococcal fibrinolysin is *staphylokinase* and for the streptococcal form *streptokinase*. The action of fibrinolysin may be to break down the protective fibrin barriers in areas of infected tissue and promote spread of the infection. Streptokinase is antigenic.

Collagenase

Several bacteria including *Bacteroides* and *Clostridium perfringens* produce this substance. The latter possesses a powerful collagenase that has a destructive effect on various tissues including muscle but its role in pathogenicity is obscure; anticollagenase antibodies neither protect against nor limit the infection.

Neuraminidase

Produced by various bacteria this is a mucinase that catalyses the hydrolysis of mucoproteins at the cell surface and may render the cell more vulnerable to bacterial attack.

Lecithinase

An alpha-toxin is produced by *Clostridium perfringens* which hydrolyses lecithin, with resulting lysis of erythrocytes and necrosis of cells.

Deoxyribonucleases (DNAses) (streptodornase)

These are important in groups A, C and G streptococci, and four distinct types, A-D, are found in *Streptococcus pyogenes*. Streptodornase and streptokinase can be used clinically to dissolve fibrinous exudates. Antibody response to DNAse B is common in patients with streptococcal infection.

Many bacteria produce a variety of substances that may or may not contribute to pathogenicity including amylases, esterases, lipases, mucinases and proteinases.

ADHERENCE

Bacterial adherence whether to smooth surfaces or to cells can be an important mechanism in the pathogenic process. Some viridans streptococci produce dextran and mutan from dietary carbohydrate and consequently can adhere to tooth surfaces; this adherence by acidogenic *Streptococcus mutans* may be the first step in the pathogenesis of dental caries. There are several examples of bacterial adherence to tissue cells and in these adherence would appear to be a prerequisite for colonisation which in turn may be a prerequisite for infection. Cholera vibrios adhere to intestinal cells, *Bordetella pertussis* to the ciliated epithelium of the upper respiratory tract and *Streptococcus pyogenes* to the pharyngeal mucosa. Lipoteichoic acid is of importance in binding certain bacteria to the membranes of epithelial cells. Some bacteria, including *Pseudomonas*, *Escherichia* and Group B streptococci possess lectin-like molecules which can bind to sugar residues, such as mannose, on epithelial cells. Fimbriae are important for the binding of certain enterobacteria.

Attachment of a virus to the cell is essential if infection is to ensue. Successful attachment requires close contact. Mucus and secretions may prevent this contact with the cell and it is suggested that the neuraminidase of influenza virus may aid infection by releasing the virus from mucus components which bind to the viral envelope and block attachment. The presence of specific receptor sites on the virus capsid or envelope and the cell membrane is also important. Thus the haemagglutinin of influenza virus binds to neuraminic acid-containing components in the membranes of cells in the respiratory epithelium. It is known that Human immunodeficiency virus (HIV) which is associated with AIDS is highly specific for T-helper lymphocytes.

FACTORS THAT PREDISPOSE TO INFECTION

Host factors

Age. Susceptibility to infection is greatest in the very young and the elderly and probably relates to the effectiveness of the inflammatory and immune responses. The proportion of patients who develop serious disease can vary. Thus poliovirus, hepatitis A, varicella and Epstein-Barr virus infections are usually mild or asymptomatic in young children, whereas in adults clinical illness is more frequent and more severe.

Genetic factors are sometimes ill-defined, as in the susceptibility of certain races to tuberculosis and the occurrence of rheumatic fever in twins, but there is increasing belief that the immune

response, which is under genetic control, is of importance. The *sex* of the host may be important; for example, the death rate from whooping cough in the first year of life is much higher in females, although exposure to the disease is equal with males. An explanation of sex differences may lie in hormonal influences.
Pre-existing disease. In burns, for example, one of the greatest threats to survival is overwhelming infection. Patients on steroids and immunosuppressive drugs and those with malignant disease are very susceptible to infection or to the reactivation of persistent infections. Tuberculosis is an example, as are herpes simplex, zoster and CMV infections. Dual or sequential infection may also be important. It is well known that viral infections of the respiratory tract such as influenza predispose to bacterial pneumonias.

Environmental factors

Nutrition has for long been known to influence infections. In recent wars infections and other diseases were rife in prisoners who were deprived of proper nutrition. Little is known about why resistance should be lowered in malnutrition although protein deficiency is known to depress the cell-mediated immune response, producing, for example, a greater susceptibility to infections such as tuberculosis, measles and herpes simplex virus. In addition, deficiencies of certain vitamins such as A and C have a deleterious effect on mucosal barriers.
Climate is important; for example, respiratory tract infections are more common in the winter months in the United Kingdom.

Occupational and social contact

Veterinary surgeons run a high risk of brucellosis, while hepatitis B is an occupational hazard of health care workers. Social factors are also of great importance and range from the standard of sanitation in a community to move individual factors such as promiscuity and parenteral drug abuse.

Local factors

Some areas of the body have very poor resistance to infection, for example the meninges, anterior chamber of the eye and joints. Other factors such as *surgical interference, poor blood supply* and the *presence of foreign bodies* all predispose to bacterial infection.

NON-SPECIFIC HOST RESPONSES TO INFECTION

There are many local and general alterations in the host in response to infection *other than* phagocytosis and the immune response.

Local responses

The inflammatory response is triggered by tissue damage at the site of bacterial or viral infection; it will be enhanced by immunological reactions. The increased blood flow and capillary permeability increase access of antibody, complement and the cells of the immune response.

Fever

This is a regular feature of systemic infection and is thought to be the product of bacterial factors such as endotoxin and host factors such as products of tissue injury, for example the endogenous pyrogens released by the leucocytes. The pathogenesis of fever is complex and may well be multifactorial including interferon and endogenous pyrogen release and antigen-antibody complex formation such as occurs late in the incubation period of systemic virus infections. Fever may have a beneficial effect by reducing the growth of some viruses and by enhancing immune responses.

Circulatory changes

Circulatory failure may be *central* or *peripheral*. Central or cardiac failure may occur as the result of damage to valves as in infective endocarditis or after damage to the mycocardium following diphtheria. Peripheral failure, or shock, characterised by events such as severe hypotension, hyperventilation, oliguria and tachycardia, can be caused by the septicaemic stage and by endotoxins.

Haematological changes

Many systemic infections produce changes in the number and proportion of white cells. In pyogenic infections a polymorphonuclear leucocytosis occurs and in typhoid fever a leucopaenia.

Immunosuppression may result from virus infection of cells of the immune system. Epstein-Barr and measles viruses infect lymphocytes and cause a variety of changes in the immune system;

as a result the skin hypersensitivity response to tuberculin may be abolished during the acute and early convalescent stages of glandular fever. Reactivity is restored when the lymphocyte populations regenerate.

Although many other changes are recognised, including muscle weakness, weight loss, electrolyte imbalance, increase in the basal metabolic rate and stimulation of the adrenal cortex, much has still to be learned about the basic physiopathological changes that occur in the host as the result of infection.

The production of symptoms and signs

The patient becomes ill at the end of the incubation period when tissue damage is sufficiently extensive. In local superficial infections this will be enhanced by the oedema and hyperaemia of the inflammatory response. In specific sites pressure may rise and pain occurs. The time taken will vary considerably from a few hours to days for staphylococcal and streptococcal infection to months for warts. Infections which spread only within the respiratory and alimentary tracts will produce symptoms after a short incubation period, thus many bacterial and viral respiratory infections and gastroenteritis are apparent within 2–3 days. Systemic viral infection—measles, rubella, polio, develop after about 2 weeks when the virus has reached target organs. The developing immune response may contribute to the systemic clinical features and enhance tissue damage as well as aiding in recovery. The clinical features may be very precisely located, as in paralytic poliomyelitis, and be poorly defined as in glandular fever or brucellosis. In some sites, for example the central nervous system, cells cannot be replaced and a permanent loss of function will ensue. At other sites, for example the liver, regeneration usually occurs with a full restoration of function.

Persistent infection

Some infections can persist for months or even years—for example tuberculosis, in which bacteria can remain in lung or renal tissue or in a lymph gland, and typhoid fever, in which the salmonellae persist in the gall-bladder. In certain instances organisms that persist can cause chronic disease; this is seen in tuberculosis and brucellosis.

However, in most infections the growth of the organism is

reduced and host responses eliminate the infection or contain the organism. Many viral infections end naturally with the elimination of the virus—for example common cold viruses and gastroenteritis viruses—although it is increasingly recognised that a number of viruses can persist after recovery. In some this may be for a limited period after the acute stage, for example poliovirus, whereas with others the virus persists for years and even for life.

The mechanisms involved are poorly understood but in some, for example herpes simplex and varicella zoster viruses, it is known that the virus can establish a latent infection in the sensory root ganglia of the spinal cord. This may be a true latent infection in that the complete virus is not present. In this way the host cell escapes immune lysis. Synthesis may be switched on and the virus reactivated to establish infection and recrudescent disease. These events occur despite the presence of antibody in the blood. In other cases the persistent infection may be quite inapparent as with cytomegalovirus and the papovaviruses. However, if the immune competence of the host is reduced, symptoms may develop as the virus multiplies and cell damage ensues. Alternative forms of persistence are seen with measles virus when the rare disease subacute sclerosing panencephalitis occurs some years after the primary infection.

FURTHER READING

Mims C A 1982 The pathogenesis of infectious disease. Academic Press
Mims C A, White D O 1984 Viral pathogenesis and immunology. Blackwell Scientific Publications, Oxford
Mims C A 1985 Virus immunology and pathogenesis. British Medical Bulletin 41. Churchill Livingstone, Edinburgh
Tobin J O'H 1983 Virus infections in the immunocompromised. In: Waterson A P (ed) Recent advances in clinical virology 3. Churchill Livingstone, Edinburgh, pp 1–18
Smith H, Pearce J H 1972 Microbial pathogenicity in man and animals. Cambridge University Press, Cambridge

3

The host response and the prevention of infections

Various mechanisms protect the individual from infection. Some are non-specific like the mucus secretion of the ciliated epithelium of the upper respiratory tract, the high acidity of the gastric juice and the vagina, whereas others such as the formation of antibodies against pathogenic organisms are quite specific.

DEFENCES IN THE HOST

Innate immunity

The determinants of this are genetically controlled in that there may be a marked difference in the response of species and individuals to infecting organisms. Man is immune to many animal infections and *vice versa*. *Salmonella typhi* produces a serious infection in man and *Salmonella typhimurium* causes only a mild localising intestinal infection, whereas in mice the situation is reversed. Within species, Algerian sheep are resistant to anthrax whereas European sheep are highly susceptible. Natural selection may play an important role in differences within species. Nutritional and hormonal factors as well as age may also have to be considered in the analysis of innate immunity.

Mechanical barriers

These will be mentioned only briefly here as they will be described in later chapters. Mechanical barriers to infection are essentially the intact skin and mucous membranes, the skin providing the more effective protection because of its outer horny layer. Joint synovia and muscle fascia also have a protective function.

Bactericidal secretions

Sebum, sweat, mucus, saliva, tears, gastric and intestinal juices act by washing microorganisms off the various body surfaces. This effect can be supplemented by other defence mechanisms, for example the action of the cilia in the upper respiratory tract. Other bactericidal substances including various acids and bacteriocines, play a useful defensive role. The components of the *normal flora* of the various anatomical sites are largely responsible for preventing the establishment of new and possibly pathogenic organisms. The mechanisms involve antibacterial substances such as lactic acid produced by the lactobacilli in the vagina, hydrogen peroxide by the viridans streptococci in the mouth and propionic acid by the propionibacteria on the skin. The normal flora plays a key role in defence against infection and when this is suppressed by the administration of broad-spectrum antibiotics pathogenic bacteria can then invade the denuded areas much more easily. *Lysozyme*, present in most tissue fluids except urine, cerebrospinal fluid and sweat, has a mucolytic effect on the cell walls of many Gram-positive bacteria; glycopeptides are split off with subsequent lysis of the bacteria.

Phagocytosis and the inflammatory response

When microorganisms successfully penetrate the body's non-specific barriers to infection and gain access to the tissues changes known as the *inflammatory response* occur. *Phagocytosis* is central to this. Phagocytes are white blood cells specialised to ingest and digest organisms. There are two types—(1) the neutrophil polymorphonuclear leucocytes (polymorphs) of the blood, the *microphages*, and (2) the large mononuclear phagocytic cells, the *macrophages*; these can either be freely circulating in the blood or fixed in the tissues, the liver sinuses (Kuppfer cells), bone marrow, lymph glands and spleen. Macrophages in the blood are called *monocytes*, those in the connective tissues *histiocytes* and those in the lymph glands *reticulum cells*. In addition to dealing with bacteria phagocytess have also a role in removing other foreign material from the bloodstream. Phagocytosis is enhanced if specific antibody to the organism and complement are present. In macrophages this is because they carry receptors for complement and the Fc portion of the immunoglobulin molecule. Macrophages can also show enhanced phagocytosis when activated by lymphokines.

When bacteria invade the body leucocytes leave the bloodstream and move towards the invaders. This is known as *chemotaxis* and operates as the result of production of chemical mediators at the site of infection. The leucocytes have then to bind to the bacteria before the organisms can be ingested and digested. Ingestion by the polymorphs occurs by an invagination of the cell membrane with the attached bacteria due to contraction of filaments in the cytoplasm. Bacteria are then enclosed in membrane-lined vacuoles in the cytoplasm of the phagocyte called *phagosomes*. Polymorphs contain granules which are *lysosomes* and these contain enzymes such as peroxidase, nucleases, phosphatases and cathepsins, as well as nucleotides, cationic proteins and lactoferrin. These lysosomes migrate towards the phagosomes containing the bacteria and fuse with them forming *phagolysosomes*. They can discharge their contents into the vacuole.

The biochemical basis for bacterial killing is not clear. H_2O_2 is formed which reacts with myeloperoxidase and disrupts cell walls—oxygen-dependent killing. The rapid fall of pH may also contribute.

Most phagocytosed bacteria are digested within 2 hours, but *Mycobacterium tuberculosis* may survive for much longer periods and *Listeria monocytogenes* and *Brucella abortus* can survive and multiply within macrophages. Survival of certain species within phagocytes may be facilitated by their secretion of substances that are deleterious or lethal to the cells.

Many viruses are inactivated by macrophages but there are infections, for example dengue, in which the monocyte is an essential host cell. Apart from their ability to destroy ingested organisms macrophages are essential for the processing and presentation of antigens to T cells to initiate the immune response. Also, macrophages can synthesise and release interferon.

Bacteria that successfully avoid or survive phagocytosis are confronted by a further line of defences, the *immune system* of the body. When the bacteria are confronted by the cells of the immune system a specific immune response occurs.

The interferon response

There are 3 types of interferon (IFN) α, β and γ: the molecular weight is approximately 20 000 (range 16–23 000) and both β and γ are glycosylated. There are at least 15 interferon genes on the host chromosome number 9. IFNs α and β are stable at pH 2. Interferon was first described as an inhibitor of virus replication but effects on cell division and the immune response (IFN γ) are now established. Synthesis of IFN α and β is induced by virus infection and some intracellular bacteria. Double-stranded RNA is a potent inducer as are synthetic polyribonucleotides. Almost all cells can make IFN if induced, although IFN αs are derived mainly from lymphocytes and IFN β from fibroblasts. IFN γ is synthesised by T cells stimulated by antigen or mitogens. Its effects include enhancement of the cytotoxic activity of NK cells, T cells and activation of macrophage cytocidal activity. Interferons show a considerable degree of species specificity in that they are most effective in the species of origin, but they can inhibit to a varying extent the replication of all viruses.

The actions of interferons are complex and several mechanisms have been described. However, there is agreement that interferons are messengers in that they are released from infected cells and attach to receptors on the cell membrane (coded for on human chromosome 21); thereafter the cells become resistant to virus infection. After exposure to IFN and double-stranded RNA, various enzyme activities are induced including an endonuclease which degrades mRNA, and a protein kinase which inactivates the peptide chain initiation factor eIF-2. Because of their broad spectrum of activity, IFNs have great potential as antiviral agents. Whatever their precise modes of action, interferon synthesis must play a part in the response to infection, both by virtue of its antiviral effects and by modifying the host immune response. There is some evidence that IFN itself may cause some of the clinical features of infection. Thus it may cause malaise and headache, fever and muscle tenderness: suppressive effects on the bone marrow are also seen during treatment with exogenous interferon.

The immune responses to infection

The specific immune responses are triggered by the foreign antigens of the invading microorganism. The responses involve a complex series of interactions between the various subsets of cells and components of the immune system. The results are important in containing and killing the pathogen thus contributing to recovery from infection, especially generalised viral infections and conferring immunity, often long-lasting, to subsequent infection. In the response to virus infections, the specific immune responses to virus-infected cells may contribute significantly to the pathological picture.

Classically the immune responses have been described as comprising two separate forms—the *humoral antibody response* and *cell-mediated* or T-lymphocyte mediated *immunity*. However it is important to realise that this is an over-simplification as T cells are involved in both responses.

The *humoral antibody response* is first detected systemically at about 10–14 days after exposure to infection. Initially the antibody is of the IgM class, but IgG appears soon after and within a few weeks is the predominant form. With generalised infections the antibody response is an important barrier to spread of virus. Antibody of the IgA class is also produced, and is present on the surfaces of the respiratory and alimentary tracts. It is an important advantage of live vaccines that they stimulate this form of protection. In some cases the antibody response declines over a period of months to years. Re-exposure to the same antigen results, within a few days, in a marked elevation of IgG levels in the blood. This is a *secondary response* and immunological memory can persist for years. A good example is the rapid rise in specific antibody in patients with herpes zoster: this is not a primary infection but a reactivation of the virus acquired many years before as chickenpox. With non-replicating antigens, the response can be accurately timed (see Figs 3.2, 3.3) but in infections the antigenic stimulation is spread over a period of time (Fig. 3.1)

Antibody molecules are produced by plasma cells or B lymphocytes derived from the liver in the fetus and from the bone marrow in the adult. Antibody synthesis occurs in response to antigen molecules presented by certain classes of macrophages. This process can be enhanced by helper T (Th) cells and modulated by

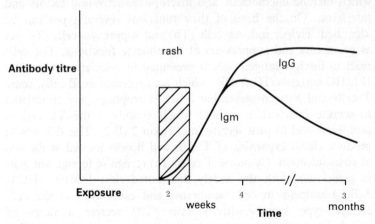

Fig. 3.1 Antibody response to a generalised viral infection with rash such as rubella

T-suppressor (Ts) cells. The antibody activity measured in the blood is a mixture of a large number of monoclonal antibodies each specific for distinct epitopes of the viral proteins. During the response, contact of the B lymphocytes with the antigens induces proliferation and expansion of the cell clone.

Antimicrobial actions of antibodies

(1) They combine with the outer components of bacteria and the viral capsid or envelope and thus (a) cause immobilisation and clumping of bacteria (b) enhance phagocytosis (c) neutralise free virus and bacterial toxins; (2) the formation of antigen-antibody complexes can activate the complement system; (3) they cause cell lysis in the presence of complement and the action of some lymphocytes (K cells), polymorphs and macrophages; (4) they modulate the expression of antigens on the surface of cells; (5) the response can be used in the diagnosis of infection (see Ch. 4).

Cell-mediated immunity (CMI)

There are several types of thymus-derived lymphocytes (T cells): they differ from B cells in that most only recognise foreign antigens on the surface of cells presented in association with 'self' major histocompatibility antigens. T cells react to antigen by the production of soluble factors—the lymphokines and interleukin which include chemotactic and macrophage-activation factors and interferon. On the basis of their functions several types can be identified. Helper-inducer cells (Th) and suppressor cells (Ts) act as promotors and suppressors of lymphocyte functions. The cells react to foreign antigen only if presented in association with class II MHC antigens (HLA-DR) which are expressed on B cells, some T cells and antigen-presenting cells. Macrophages are stimulated to secrete interleukin 1 (Il-1) which stimulates the Th cell to proliferate and in turn secrete interleukin 2 (Il-2). The Il-2 acts to produce clonal expansion of T cells and B cells located at the site of virus infection. Cytotoxic T cells (Tc) recognise foreign antigens in association with the widely distributed class I MHC (HCA-A,B,C) antigens in cell membranes and cause lysis of the cell. Delayed-type hypersensitivity cells (Td) secrete a range of lymphokines which augment the immune response. Several other types of T cells are known, including the natural killer (NK) cells.

These cells are immunologically non-specific and can kill virus-infected cells: their activity is increased by interferon γ.

In some virus infections, growth of the virus in cells of the immune system will suppress aspects of the host response and may predispose to infections. The human T-lymphotropic virus type III grows in and destroys Th cells. As a result, in the later stages of the acquired immune deficiency syndrome (AIDS) patients suffer from a wide spectrum of infections.

Recovery from virus infection is achieved mainly by the complex series of cell-mediated reactions, including the various T cells, macrophages and NK cells and interferons. Of these, macrophage activity and interferon will be effective soon after infection. The T cell responses will build up over a few days. Antibody is not of such importance, although its time of appearance in systemic viral infections indicates that it will assist in recovery. Clinically, children with congenital defects of cell-mediated responses fare badly with respect to viral infections, although poliovirus is a notable exception. The ability to produce an antibody response is important in resistance to bacterial infection.

Immunopathology

The immune response to infection, especially to viral infection, may contribute significantly to cell and tissue damage. There are some animal models, for example lymphocytic choriomeningitis virus in mice, in which acute disease and death of the infected animal only occur if the immune system is intact. In addition to the mechanisms described above the formation of immune complexes is a significant cause of pathology by activation of the complement system, aggregation of platelets and thus enhancement of inflammation and cellular infiltration. Such processes are important in the pathogenesis of viral skin rashes and soluble complexes can be deposited on the walls of small blood vessels and renal glomeruli. This can result in rash, polyarteritis nodosa and glomerulonephritis in hepatitis B infection. In extreme cases antigen-antibody complex formation can lead to disseminated intravascular coagulation. Delayed-type hypersensitivity cells (Td) are important in the evolution of the inflammatory process.

In infections with Mycobacteria and *Brucella* spp there is reason to believe that cellular immunity of a delayed hypersensitivity type is the primary basis of acquired immunity. This can be highlighted in children with agammaglobulinaemia or hypogammaglobulinaemia. Although they may be resistant to infections with these intracellular bacteria, their resistance to infections with other organisms is considerably reduced. CMI is detectable in several infections by skin testing.

Acquired immunity

$$\text{Active} \begin{bmatrix} \text{Natural} \\ \text{Artificial} \end{bmatrix}$$

$$\text{Passive} \begin{bmatrix} \text{Natural} \\ \text{Artificial} \end{bmatrix}$$

Active natural immunity

This results either from a clinical or subclinical infection and is directed towards the organism that causes the infection. Some infections such as whooping cough, diphtheria and most systemic viral infections such as measles produce long-lasting immunity, whereas in others such as gonorrhoea the period of immunity may be quite brief. It is worth noting that an attack of tetanus does not produce any immunity and an individual who has suffered the disease must be actively immunised to prevent future attacks.

Active artificial immunity

This results from immunisation procedures. It can be produced by injection of living attenuated organisms as in vaccination against tuberculosis, rubella and measles, by injection of killed organisms as in protection against whooping cough, the enteric fevers, rabies and hepatitis B, and by administration of toxoid as used for the prophylaxis of tetanus and diphtheria (see Figs 3.2 and 3.3). Subunit vaccines are available to protect against influenza virus: the preparations contain the surface antigens of the viral envelope. The immunogenicity of any non-living vaccine can be enhanced by the form of presentation of the antigen, for example, by adsorption to adjuvants such as alum.

PREVENTION OF INFECTIONS 37

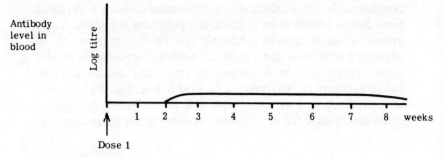

Fig. 3.2 Primary response

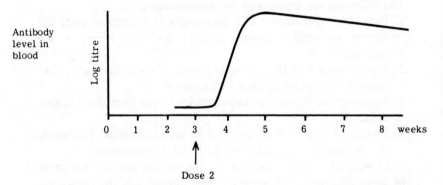

Fig. 3.3 Secondary response

Passive natural immunity

This is produced in the fetus by transplacental IgG antibodies from the mother and means that the baby will have some form of minimal protection to certain infections for the first two or three months of life. Antibodies of the IgA class are also acquired in the first few weeks from the mother's milk.

Passive artificial immunity

Administration of immune serum which may be either human or prepared in another species such as the horse, confers this form of transient immunity for up to 6 months. There are dangers in giving heterologous antiserum because of the possibility of severe hypersensitivity reactions. Human immunoglobulin does not produce such effects and is safe to use; pooled human immunoglobulin will be sufficient to protect against infections prevalent in the donor population. In other infections, hyperimmune globulin is prepared from donors known to be immune to a particular organism. Active immunisation is always preferable to passive because of its prolonged protection and comparative safety. Nevertheless passive immunisation may be important in preventing disease if given before exposure or may prevent or modify infection if given soon after exposure. It may soon be possible to use human monoclonal antibodies specific for the important epitopes of the microorganism.

IMMUNISATION

The following are indications for immunisation:
1. Protection of members of a community in infancy or early life against potentially serious diseases that can occur in the community
2. Protection of travellers going to countries where there are infections not prevalent in their own country
3. Protection of nations or communities in the face of a possible epidemic
4. Protection of groups at special risk, such as hospital personnel, the old and infirm, and those with debilitating diseases

As with all therapy, the risks of the vaccine and its cost must be balanced against the morbidity and mortality of the disease. It is important that the responses induced by immunisation should be long-lasting and should mimic the response to the natural disease as closely as possible.

Vaccines

There are four forms of antigen used in immunisation procedures: (1) *Attenuated living*; (2) *Killed*, by chemicals such as formalin or alcohol; (3) *Toxoids*, toxins that have been made non-toxic either by the action of heat or the addition of formalin; (4) *Subunit*, prepared by purifying the important antigens of the microorganism.

There are a number of problems associated with immunisation: some of the specific complications are listed below. In general, live vaccines should not be given in pregnancy due to the risk of fetal infection and should be avoided in compromised patients who show an increased rate of severe complications.

Immunisation schedules (see Table 3.1)

PREVENTION OF INFECTIONS 39

Table 3.1 Schedule of vaccination and immunisation procedures—1972—extracted with minor amendments from Memo. of Standing Medical Advisory Committee (HMSO, July, 1972)

Age	Vaccine	Interval	Notes
During the first year of life	Diph/Tet/Pert and oral polio vaccine. (First dose)		The earliest age at which the first dose should be given is 3 months, but a better general immunological response can be expected if the first dose is delayed to 6 months
	Diph/Tet/Pert and oral polio vaccine. (Second dose)	Preferably after an interval of 6 to 8 weeks.	
	Diph/Tet/Pert and oral polio vaccine. (Third dose)	Preferably after an interval of 4 to 6 months.	
During the second year of life	Measles vaccine	After an interval of not less than 3 weeks after any other live vaccine.	Although measles vaccination can be given in the second year of life, delay until the age of 3 years or more will reduce the risk of occasional severe reactions to the vaccine.
At 5 years of age or school entry	Diph/Tet and oral polio vaccine or Diph/Tet/Polio vaccine.		These may be given, if desired, at 3 years of age to children entering nursery schools, attending day nurseries or living in children's homes.
Between 10 and 13 years of age	BCG vaccine		For tuberculin-negative children.
All girls aged 11 to 13 years	Rubella vaccine	There should be an interval of not less than 3 weeks between BCG and rubella vaccination.	All girls of this age should be offered rubella vaccine whether or not there is a past history of an attack of rubella.
At 15 to 19 years of age or on leaving school	Polio vaccine (oral or inactivated) and tetanus toxoid		

Complications of immunisation procedures

Whooping cough vaccine

Much controversy has surrounded the prophylaxis of whooping cough because of the possibilities of encephalopathy and convulsions, and an expert committee was set up by the Department of Health and Social Services in the United Kingdom to study all aspects of whooping cough and its prophylaxis. A report was published in 1977 and some of the conclusions are as follows:

1. Whooping cough remains a serious disease in young children and 10% of those infected under 2 years of age have to be admitted to hospital.
2. Protective efficacy of the vaccine has been fully restored in recent years by changes in its antigens. The prevalent serotype 1, 3, has now been added to the other two serotypes 1, 2, and 1, 2, 3.
3. Many symptoms and signs purported to result from vaccination are not specific to pertussis vaccine, and retrospective analysis of cases of reported adverse reactions does not permit conditions such as childhood convulsions or encephalopathy that occur coincidentally and are unrelated to immunisation to be differentiated from those in which a relationship appears to exist.
4. A full course of immunisation with vaccine not only reduces the attack rate of the disease but also the severity of the illness.
5. Risks with current vaccines are slight, but care must be taken not to administer vaccines with a pertussis component where there are parental objections or medical contra-indications such as pre-existing neurological conditions. (Diphtheria and tetanus components should still be given.)
6. The first dose of triple vaccine should be given at 3 months of age so that more infants will be protected at the vulnerable period. Also at this earlier age there is a lower risk of febrile convulsions.

Typhoid and paratyphoid vaccine

Fever rigors and headache are common complications.

Polio vaccine

This live vaccine (Sabin) can be excreted for some weeks after administration and may be spread to other children or adults.

When young children are immunised it is important to ensure that their parents are also protected. The vaccine has not been shown to cause damage to the fetus, but if protection is essential in pregnancy the killed vaccine (Salk) can be given. There is a significant increase in the risk of CNS infection if live vaccine is given to immunocompromised patients who lack an effective humoral antibody response.

Rubella vaccine

Should not be given in pregnancy as fetal infection has been demonstrated. Antenatal screening is undertaken to detect seronegative women who are then immunised in the puerperium: a further pregnancy should be avoided for 3 months.

Measles vaccine

Contains attenuated virus. It must not be given in pregnancy or to the immunocompromised patient.

Although not listed in the schedule, the following vaccines are effective and available.

Mumps vaccine—an attenuated vaccine. *Yellow fever*—a live vaccine available in this country for travellers to endemic areas. *Rabies*—the human diploid cell vaccine is much safer and more antigenic than the earlier vaccines: it can be given to protect staff in quarantine areas and to those at risk occupationally. It is also used in conjunction with passive immunisation in the post-exposure treatment of animal bites. *Hepatitis B*—this vaccine contains the small particles of the hepatitis B surface antigen purified from the plasma of carriers of the virus. Three doses of the vaccine are given at intervals of 1 month and 6 months. Apart from local effects at the site of injection, no serious complications have been recorded.

Influenza—severe generalised reactions may occur and local reactions are common especially to the less purified preparations. Vaccines prepared by disrupting the virion and purifying the envelope components are much less reactive. Attenuated live vaccines have been prepared by selection of temperature-sensitive (ts) mutants of the virus. Alternatively, attenuated parental strains are used to prepare new vaccines by exchanging their genes for the surface antigens (H & N) for those of the new strain. Live vaccines are not widely available as yet. It is important to remember that the antigenic composition of the vaccine must be varied as the virus changes. In the United Kingdom vaccines contain two influenza A and one influenza B components.

Future developments

Modern techniques of DNA cloning and expression in *Escherichia coli*, yeasts and cultured cells have made it possible to prepare specific gene products for immunisation. A vaccine for hepatitis B has been produced in yeasts and many more will follow.

Chemically-defined antigen preparations will be available in future instead of many of the relatively crude preparations now in use. Thus it is possible to identify the specific capsid polypeptide of importance in poliovirus neutralisation. Such antigens if available in sufficient quantity may be combined for administration in polyvalent vaccines. Another interesting prospect has been raised by the incorporation of the hepatitis B surface antigen gene in the DNA of vaccinia virus. This modified virus can infect the skin of experimental animals and, as a result, induces immunity to smallpox and the hepatitis B virus. Several antigens could be administered in this way: vaccinia in its present state of attenuation is not suitable for human use but a modified strain could be.

STERILISATION

Sterilisation is an absolute term and denotes the complete removal of all microorganisms from an object. There are several physical methods of sterilisation.

Heat

Dry heat and *moist heat* can be used.

Dry heat

This kills by causing oxidation of the bacterial cytoplasm and has the advantage of speed and of being easily controllable. Its use is restricted because certain articles, plastics, rubber goods and fabrics, cannot withstand high temperatures.

Simple flaming. This can be used for bacteriological loops and the points of forceps. Incineration is used for the disposal of contaminated carcasses, plastics and dressings.

Hot-air oven. This is a commonly used dry-heat steriliser. It is heated by electricity and has a fan to ensure even distribution of air. It is used for sterilising glassware, including glass syringes, powders, oils and greases that are impenetrable to moisture, as well as cotton wool pledgets and paraffin gauze dressings. Articles should be packed so as to allow circulation of the air. The *sterilisation time* is the sum of the *heat penetration time*, the *holding time*, the time required to sterilise a load when the required temperature has been reached, and the *safety time*, which is usually half the holding time. Glassware is generally sterilised by exposure to 160°C for one hour, or to 180°C for 20–30 minutes (the holding plus safety times) but these times may have to be doubled when powders, oils or greases are sterilised. The sterilisation cycle can take up to three hours including heat penetration and cooling times.

Conveyor oven. In a conveyor oven the method of sterilisation uses infra red radiation; the temperature of operation is 180°C for 10–15 minutes. This is used for large-scale sterilisation of glass syringes.

Moist heat

This kills by coagulation of the bacterial cytoplasm and is effective at lower temperatures than dry heat.

At temperatures below 100°C. Pasteurisation kills non-sporing pathogens. In the pasteurisation of milk two methods may be used: 64–66°C for 30 minutes (the *'holder'* method), or 72°C for 15–20 seconds (the *'flash'* method). Some vaccines are also heated at lower temperatures such as 60°C for 1 hour or longer to kill the micro-organisms but preserve antigenicity.

At 100°C (boiling).
Boiling at 100°C for 5–10 minutes kills most vegetative organisms, viruses and some, but not all, spore-forming organisms; it therefore does not ensure sterility. Boiling may be used for disinfecting contaminated crockery or cutlery.

At 100°C (steam). A temperature of 100°C is produced by pure steam in equilibrium with water boiling at normal atmospheric pressure and is used to sterilise substances such as bacteriological culture media that would be damaged by autoclaving. Steaming may be done either with a single exposure for 90 minutes, or with intermittent exposures, 20–30 minutes on 3 consecutive days (*Tyndallisation*). The principle underlying this practice is that the first steaming kills vegetative organisms and any spore-forming organisms present germinate before the next heating. These are then killed off during the next exposure to steam.

At temperatures above 100°C. The efficiency of steam under increased pressure as a *sterilising agent* is due to its condensation into a small volume of water when it meets the cooler surfaces of articles to be sterilised, liberating considerable latent heat to these surfaces. This is the principle involved in the *pressure cooker* and *autoclave*.

The autoclave

Two main groups of autoclaves are recognised; those used for porous loads such as dressings and preset operating theatre trays in hospitals, and those used as laboratory and instrument sterilisers.

The sterilisation cycle

This starts when the door of the autoclave is closed and ends with removal of the load and is as follows:
1. Removal of air from chamber
2. Sterilisation of the load
3. Exhaustion of steam from chamber after sterilisation
4. Driving and cooling of load

Removal of air from chamber

Air must be removed from the chamber for several reasons: (1) the presence of air seriously lowers the temperature at a given pressure; (2) air hinders the penetration of steam into porous materials such as surgical dressings; and (3) because air is denser than steam it forms a layer at the foot of the chamber of the autoclave and produces a lower temperature there.

Minimum exposure times for sterilisation of instruments are:

4 minutes at 134°C (at 221 kPa)
15 minutes at 121°C (at 104 kPa)
30 minutes at 115°C (at 69 kPa)

Design of autoclaves

Design differences depend on the method of removing air from the chamber; there are two types in general use.

Downward displacement sterilisers. These are used in laboratories and as instrument sterilisers. Steam enters near the top of the chamber and the air is displaced downwards through the load and the discharge channel at the bottom of the chamber. This channel is fitted with a thermostatic steam trap that remains open until all the air and condensate has been removed. Once pure steam only is present the temperature of the discharge rises to 121°C when the trap automatically closes and prevents further escape. Steam also enters the jacket independently and facilitates drying of the load. Once the sterilising cycle is completed the steam supply is stopped and filtered air enters the chamber to cool and dry the load.

High pre-vacuum sterilisers. Used mostly in hospitals these are equipped with electrically-driven pumps that can remove virtually all the air from the chamber and this high vacuum ensures that steam penetrates the load rapidly. The effect of these sterilisers is to cut the sterilisation time; temperatures are higher (134°C) and pressures are greater (221 kPa gauge pressure). They are particularly useful when heat-sensitive materials have to be sterilised, because of the short sterilisation time.

Tests for efficiency of heat sterilisation

Chemical indicators

These are convenient because they give an immediate result, and may be placed inside or outside the load. Examples of such are *Browne's tubes*, small sealed tubes containing a red solution that turns to green when an adequate temperature has been reached.

The Bowie-Dick autoclave tape test

This measures temperatures in packs and indicates the degree of steam penetration and air removal. A sheet of paper with a diagonal cross of autoclave tape is placed in a standard test pack, and a uniform colour change throughout the diagonal tape within a specified time indicates successful sterilisation.

Electrical indicators

Thermocouples are used to measure temperatures in loads in different areas of the steriliser. They are electrical leads that are connected to a recording instrument outside the steriliser which indicate the temperature reached during the autoclaving process.

Spore indicators

A preparation of bacterial spores such as *Bacillus stearothermophilus* is placed in a load and after autoclaving is tested for viability by transfer to another culture medium. It indicates the conditions of the load where it was placed but gives no indication of the sterilising performance in other loads.

Radiation

Two types of radiation are used, *non-ionising* and *ionising*.

1. Non-ionising. Bright sunlight can kill bacteria and viruses because of its ultraviolet rays. The effective rays are those with wavelengths of 24 000–28 000 nm (2400–2800 Å). Ultraviolet lamps emit rays within the above wavelengths but the value of these in sterilisation is very restricted because of their extremely poor penetrating power. They may be used for disinfection of work surfaces but spores are resistant to the rays.

2. Ionising. Ionising radiations are highly lethal. They may be either high energy electrons (*cathode rays*) from a machine such as a linear accelerator or *gamma rays* from radioactive isotopes such as cobalt-60. The mode of action of such radiations is to ionise vital cell components such as DNA, and a dose of 2.5 Mrad (25 Gy) is generally adequate for this.

Articles that may be sterilised by radiation include disposable rubber gloves, needles and syringes, other plastic equipment and sutures. Advantages of this form of sterilisation are its efficiency, negligible rise in temperature and the fact that articles can be packed and sealed before sterilisation. Amongst the disadvantages are that textiles lose their tensile strength and colour, chemical changes occur in some products, and glassware darkens; in addition the sterilisation time is long.

Filtration

Various types of filter are used to remove organisms from fluids or from the air. Bacteria are trapped by direct impaction with the fibre or by electrostatic precipitation. Viruses, rickettsiae, chlamydias and mycoplasmas will not be retained by filters unless a suitable pore size is used.

Fluids

Fluids and bacterial cultures can be rendered free from bacteria by using filters with a pore size of around 0.75 μm. Various types have been used including *Chamberland* (unglazed porcelain), *Berkefeld* (diatomaceous earth) and *Seitz* (asbestos). Membrane filters of cellulose acetate and nitrate (*Millipore* and *Gradocol*) are now widely used and are available with specific pore sizes.

Air

Air filtration is essential for operating theatres and isolation units; filters incorporate fibrous material such as glass wool, asbestos fibres in sheets, or gauze. Electrostatic precipitators are useful in trapping and removing particles from the air.

Gas

The most widely used gases are *ethylene oxide* and *formaldehyde*. They act by alkylating the $-SH$, $-OH$, $-COOH$ or $-NH_2$ groups of proteins and nucleic acids under closely defined conditions of temperature and humidity. *Ethylene oxide* is used in hospitals and industrial processes but because of its explosive properties is unsuitable for fumigating rooms. It has a potent sporicidal action but is also toxic for man. It can diffuse easily through paper, plastics and fabrics and can be used for sterilising the following: swabs, electrical and dental equipment, components of the heart-lung machine, endoscopes, ureteric and cardiac catheters.

Formaldehyde

This gas is also highly lethal to organisms and spores and can be used in aqueous or gaseous forms. It is used for fumigating rooms and for disinfecting mattresses and bedding.

The advantages of sterilisation by gas lie in the fact that most articles are not damaged in the process and that penetration into porous materials is usually good. However, this gas is highly irritant.

DISINFECTION

Disinfection is not an absolute term like sterilisation, because although it implies destruction of bacteria it does not guarantee such an effect on spores or viruses.

Antisepsis. Antisepsis is the disinfection of living tissues and antiseptics are the milder disinfectants that can be allowed to come in contact with living tissue without causing harm.

Factors influencing the action of disinfectants

The action of disinfectants include protein and lipid denaturation in the cytoplasmic membrane, the blocking of biosynthetic pathways and damage to energy-yielding systems in the cell. Several factors influence their action:

Organisms

Most vegetative bacteria except acid-fast bacilli are killed under optimum conditions for disinfectant action, although spores and some viruses are highly resistant. 2% aqueous glutaraldehyde has proved effective against spores; viruses are also sensitive to this as well as to formaldehyde, iodine and hypochlorite.

Rate of action

Some disinfectants act quickly, others slowly, depending on the number of organisms, the amount of organic matter present, the temperature and pH of the environment and the concentration of the disinfectant.

Number of organisms

In general the larger the numbers of organisms the longer the period of exposure required, and it should be remembered that bacteria may number many millions in situations where disinfectants are used.

Condition of organisms

Moisture in and around organisms is essential for disinfection.

Organic matter

Organic matter such as pus, blood and excreta, block penetration by coating the organisms; in addition, if there is much liquid organic matter this may serve to dilute and actually inactivate the disinfecting agents.

Concentration

The rate of killing of bacteria varies directly with the concentration of the disinfectant, and the failure of disinfection is often the result of failure to use recommended concentrations.

Temperature

A rise in temperature usually increases the rate of disinfection and disinfectants are more effective made up in hot water.

pH

This is important in the use of many disinfectants. An agent such as glutaraldehyde functions only in alkaline solutions, whereas compounds such as phenol work optimally in an acid medium.

Volume

Usually a greater volume of disinfectant produces more effective sterilisation as the effect of any inactivators present is reduced.

Formulation

Some disinfectants such as chlorhexidine operate best in an aqueous solution, or with 70% alcohol. Some are combined with detergents in an attempt to achieve simultaneous disinfection and cleaning.

Deterioration

Many substances deteriorate when diluted with water and freshly-prepared solutions must be used. The longer the period the disinfectant is kept the less effective it will become, and bacteria have been known to survive and multiply in ageing disinfectants.

Inactivation

All disinfectants may be inactivated by certain materials. For example, hard water and soaps inactivate quaternary ammonium compounds and rubber inactivates phenols and chlorhexidine. Long storage and exposure to air will cause the disinfectant to deteriorate.

Evaluation of disinfectants

1. Manufacturers' tests

Disinfectants are frequently tested under conditions that may not be relevant to later 'in-use' situations. Organic matter and inactivating materials may be absent, large volumes and high concentrations of the disinfectant may be used and the challenge may consist of only small numbers of bacteria. The *Rideal-Walker* test is a standard test and has been used for many years. It compares dilutions of the test disinfectant with known concentrations of phenol in the killing of *Salmonella typhi*. From the results a phenol coefficient is calculated; the higher the number the better the performance of the disinfectant. The calculation of the phenol coefficient is particularly *unhelpful* because (1) comparison with phenol has no meaning for any other type of disinfectant; (2) the conditions are not related to practical use (no organic matter is included for example); (3) only one species of organism is used; and (4) minor variations in technique can produce wide variations in the final result. Other phenol coefficient tests are the *AOAC* (Association of Official Agricultural Chemists) test which uses both *Salmonella typhi* and *Staphylococcus aureus*, and the *Chick-Martin* test which is a slight improvement in that it incorporates organic matter. It is only acceptable as a test for *phenolic* disinfectants however.

2. Reference laboratory tests

Tests carried out by the Disinfection Reference Laboratory of the Public Health Laboratory Service in London include the *Kelsey-Sykes* test which was introduced in 1969. It is suitable for all disinfectants and uses standard test strains of *Staphylococcus aureus*, *Pseudomonas aeruginosa*, *Proteus vulgaris* and *Escherichia coli*, grown on a chemically-defined medium; yeast cells are incorporated as a source of organic material.

3. In-use tests

Such tests assess the activity of a disinfectant in solutions that are actually in use in hospitals. They are much more realistic in that they take into account factors and circumstances that affect the performance of the disinfectant in a real-life situation.

CHEMICAL DISINFECTANTS

Alcohols

Examples

Isopropyl alcohol and ethyl alcohol.

Use

They are mainly used in 70% concentrations diluted with water and are useful as skin disinfectants before injections or venepuncture; they are also used for clean surfaces such as trolley tops.

Aldehydes

Examples

Formaldehyde
Glutaraldehyde

Use

Gaseous formaldehyde is used for fumigation: glutaraldehyde is useful for disinfection of clean instruments such as endoscopes but it is not suitable for use on dirty surfaces. It is irritant particularly to the eyes. A 2% solution of glutaraldehyde is also virucidal.

Diguanides

Examples

Chlorhexidine (Hibitane)
Picloxidine

Use

Chlorhexidine is a useful disinfectant for skin and mucous membranes and is used as a 0.5% concentration in 70% alcohol. It is recommended as suitable for skin preparation of the patient, before sternal marrow aspiration, lumbar puncture and before surgery. It is not used for disinfection of instruments, equipment and surfaces because it is inactivated by many materials. It is less irritant than formaldehyde and is a useful agent for hospital use. Hibiscrub (4% chlorhexidine in detergent) is a useful pre-operative scrub for surgeons' hands.

Halogens

Examples

Chlorine
Iodine
Iodophors: Povidone-iodine
            Betadine
            Pevidine
            Wescodyne
Hypochlorites: Chloros
               Milton
               Domestos
               Eusol

Use

Chlorine is an effective disinfectant of water and *iodine* with alcohol is a powerful skin disinfectant which is recommended for use on patients before surgery. *Iodophors* contain iodine with surface active non-ionic detergents and are effective skin disinfectants because unlike iodine they do not stain or irritate the skin. They are sporicidal. *Hypochlorites*, derivatives of chlorine, are widely used for disinfecting dairy equipment and feeding bottles and are used in laboratory discard jars. Eusol is a solution of hypochlorites useful for wound cleansing. Hypochlorites are seriously inactivated by organic material, for example remains of milk in inadequately-cleaned feeding bottles and have a corrosive action on metals; they are, however, very effective in decontaminating floors and other surfaces.

Phenolic compounds

Examples

Phenol (carbolic acid)

Cresol
Chloroxylenol (Dettol)
Clear soluble phenolic fluids: Lysol
Sudol
Hycolin
White phenolic fluids (Izal)
Black phenolic fluids (Jeyes Fluid)
Hexachlorophane (pHisoHex)

Use

Phenolic compounds are widely-used disinfectants, although phenol itself is seldom if ever now used because of its irritant effect. The clear soluble fluids, white fluids and black fluids constitute *coal tar disinfectants* and are the basis of many proprietary disinfectants. They are widely used in disinfecting surfaces, in general sanitation and in bacteriological laboratory discard jars. *Hexachlorophane* can be incorporated in detergents and soaps, and as pHisoHex is a useful skin disinfectant for a pre-operative surgical scrub; it is retained on the skin for long periods and the antibacterial effect is slowly cumulative. There are possible hazards in using pHisoHex on infants as absorption with subsequent toxic effects may occur.

Quaternary ammonium compounds (QACs)

Examples

Cetrimide (Cetavlon)
Benzalkonium chloride (Roccal)
Cetrimide + chlorhexidine (Savlon)
Cetrimide + picloxydine (Resiguard)

Use

QACs are cationic detergents; the detergent effect on surface activity is due to an organic cation in solution. These *QACs* or *QACs + diguanides* have a powerful antibacterial activity and are widely used for wound and skin disinfection, for instrument disinfection and environmental sanitation.

Ampholytes

Example

Tego compounds

Use

These amphoteric compounds combine a detergent and bactericidal action and may be used in dairy and meat industries for environmental sanitation and application to surfaces.

SUMMARY OF MEDICAL APPLICATIONS OF CHEMICAL ANTISEPTICS AND DISINFECTANTS

Disinfection of skin

Mechanical removal of bacteria

A surgical scrub or shaving the patient's skin will reduce the skin flora but are not sufficient on their own. Chemical disinfection of the skin is essential although sterilisation is impossible. Desirable properties of skin disinfectants include rapidity and persistence of action with a bactericidal effect over a wide range of organisms; they should be non-irritant. Few agents meet these requirements.

Patient's skin

Before an injection
—Soap and water
—70% ethyl or isopropyl alcohol with the addition of 1–2% iodine or 0.5–1% chlorhexidine.

Before surgery
—Quaternary ammonium compounds (cetrimide)
—Alcoholic solution of iodine
—Alcoholic solution of chlorhexidine
—Chlorhexidine-cetrimide mixture (Savlon)
—Povidone-iodine solution

Surgeons' hands

—Soap and water with a Savlon or alcoholic chlorhexidine rinse.
—Hexachlorophane surgical scrub

—Povidone-iodine surgical scrub
—Hibiscrub
These measures can also be used by nurses changing dressings.

Disinfection of clean, inanimate surfaces (trolley, furniture)
—Hypochlorite with detergent
—70% ethyl or isopropyl alcohol in water
—0.1% aqueous solution of chlorhexidine
—0.1% aqueous solution of cetrimide or benzalkonium chloride.
—1% hypochlorite
—2% glutaraldehyde

Disinfection of dirty surfaces (floors, toilets)

Phenolic compounds such as Lysol and Sudol

Disinfection of clean instruments and other apparatus
Buffered glutaraldehyde fluid
Ethylene oxide gas

Protection of health workers

Specific immunisations should be implemented in staff, the range depending on the nature of the work to be undertaken. However, a high standard of personal hygiene is essential in staff and they must remember that patients' blood and secretions may be infectious. This does not mean that all patients require full barrier nursing, but rather that unnecessary contamination of skin and mucous membranes must be avoided. There are defined procedures for the care of patients in infectious disease units; within general medical and surgical wards the dissemination of organisms from infected patients must be controlled to protect other patients and staff. Particular care is needed in the management of carriers of the hepatitis B virus and those at risk of AIDS, especially when invasive procedures are necessary. High risk guidelines should be available for the care of such patients in the wards and in operating theatres and will include instructions for the safe disposal of contaminated equipment, instruments and linen. It is important to avoid accidental needle pricks or cuts with used instruments.

It must be remembered that laboratory staff may also be exposed to infection from blood and other specimens submitted for examination. It is the responsibility of the clinician to inform the laboratory of any special known or suspected infection risks so that appropriate precautions may be taken.

FURTHER READING

Department of Health and Social Security 1977 Whooping cough vaccination. HMSO, London

Lowbury E J L, Ayliffe G A J, Geddes A M, Williams J D 1975 Control of hospital infection. Chapman and Hall, London

Maurer I M 1974 Hospital hygiene. Arnold, London

Mims C A 1982 The pathogenesis of infectious disease. Academic Press

Mims C A 1985 Virus immunology and pathogenesis. British Medical Bulletin 41 Churchill Livingstone, Edinburgh

Raynaud M, Alouf J E 1970 Microbiol toxins 1: 67

Rubbo S D, Gardner J F 1965 A review of sterilisation and disinfection. Lloyd-Luke, London

Sykes G 1972 Disinfection and sterilisation. Chapman and Hall, London

Thompson R A 1974 The practice of clinical immunology. Arnold, London

Weir D M 1983 Immunology. An outline for students of medicine and biology. Churchill Livingstone, Edinburgh

4

Diagnosis of infections

LIAISON WITH THE LABORATORY
Liaison between clinical and laboratory staff is of paramount importance in the investigation of a patient's illness and whenever there is the possibility of infection the bacteriologist, virologist or senior technician should be consulted about the laboratory investigations that should be undertaken. It is important for the clinician to give *relevant* information to the laboratory on the request form accompanying each specimen. Examples of information that may be important include the presentation of the patient's illness and the tentative clinical diagnosis; how long the person has been ill; whether he or she has been abroad recently; the relevant immunisation history and whether any antimicrobial therapy was given to the patient before the specimen was taken. For all specimens the date and hour of taking the specimen must be indicated. Suspected or known high risk infection must be indicated.

The sooner a specimen reaches the laboratory the more reliable are the results that can be obtained and if specimens are important or urgent the doctor or nurse should deliver them personally. Swabs collected for virus isolation *must* be placed in a virological transport medium, and if a delay of more than 2 hours is anticipated they should be *refrigerated*. Specimens for the isolation of chlamydiae must be stored at −60 °C or lower if transport to the laboratory is not immediately possible.

Important and relevant laboratory findings will be reported immediately by most bacteriologists either in person or by telephone to the ward staff and before the typewritten report is despatched.

NOTES ON THE COLLECTION OF SPECIMENS
Remember to ensure that specimen containers are properly sealed

and that the outsides of the containers are not contaminated. Request forms must *not* be wrapped around the container. Use self-sealing plastic bags, ensure that request form and specimen are correctly matched and use the correct procedure for 'Risk of Infection' specimens. In the following sections, the instructions cover the collection and transport of specimens for both bacteriological and virological examination.

Swabs (see Fig. 4.1)

Numbers of organisms isolated from swabs in the laboratory may bear no relation to the numbers in the original clinical lesion. This is due to physical factors such as desiccation, to bacterial overgrowth and to the toxic effects of the metabolism of certain bacteria. In the laboratory there is a loss of over 90% of the organisms when swabs are plated out directly because many are retained on the swab. Transport media may be used in an attempt to maintain viability if the period of transit is to be long. Virus transport medium is essential for swabs and scrapes of exudate while special transport medium is needed for chlamydiae.

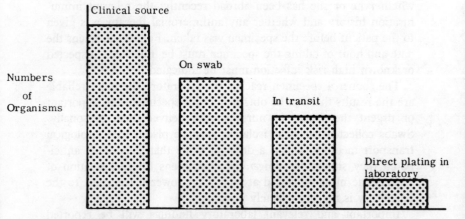

Fig. 4.1 Progressive loss of organisms from clinical source to laboratory

Blood for culture

Take blood before any chemotherapy, preferably 3–4 paired cultures per day at intervals. (Paired cultures are useful in checking for contaminants). Blood can be taken without waiting for a rigor. The site of venepuncture must be adequately cleansed (see Ch. 3).

Use a bottle containing penicillinase if the patient is on pencillin or ampicillin. Most laboratories have an incubator available for blood cultures taken outwith laboratory hours. When blood is taken the needle should be changed before inoculating the broth medium; this will minimise the risk of contamination.

CSF

Place 2–3 ml in a sterile container. Contact the bacteriologist if meningitis is suspected.

Cervical swab or scrape

Cervical swabs and scrapes are collected from patients with suspected or overt cervical infection and also from patients with infections of the uterus, salpingitis and pelvic inflammatory disease. Take under direct vision (use a speculum) and in good light. In gonococcal infection plates are best inoculated at the bedside. Special transport media must be used for the isolation of viruses and chlamydiae.

Cough plate (for whooping cough)

Plates are available from the laboratory on request. Expose to several coughs and return immediately to laboratory. However, pernasal swabs are more effective.

Eye swabs

Most causes of conjunctivitis are non-bacterial and special transport media for viruses and chlamydiae are available. Smears from swabs or scrapes are made in the ward and dried in air; do not fix if for virology; if immunofluorescent examination is required slides should be fixed in acetone.

Faeces

The specimen is most easily collected from adults by having the patient pass faeces on to a wad of clean toilet paper. A 'spoon' is provided with the container. Send enough material to occupy a third of a universal container. A 'hot stool' is required for examination for cysts, ova and motile parasites.

Peritoneal dialysis fluids

Send in a dry sterile universal container.

Peritoneal, pericardial, pleural and joint fluids

Send in a special sterile bottle containing anticoagulant.

Pernasal swabs (for whooping cough)

Use a special, soft, wire-mounted swab. Pass it along the floor of the nasal cavity to the posterior wall of the naso-pharynx and rotate the swab gently. Nasopharyngeal secretions are suitable for the direct detection of viral infections of young children. A mucus extractor with a fine catheter should be used to collect the specimen.

Pus

Whenever possible aspirate pus and inflammatory exudates using a sterile syringe and needle. Send in sterile bijou or universal container.

Rectal swabs

These are inferior to a specimen of faeces.

Serological tests

5 ml of clotted blood should be despatched in a dry tube. 10 ml will be necessary if many tests are required. Paired samples are mandatory, an *acute phase* sample followed by a *convalescent phase* sample 10–14 days later. Fourfold or greater rises in antibody titres and presence of specific IgM are of diagnostic significance.

Skin

Scrapings should be collected into black paper for the diagnosis of fungal infections. The core of a molluscum contagiosum lesion is sent in a dry sterile container. Scrapings from the bases of lesions can be sent in virus transport medium. Vesicle fluid can be collected by filling a fine syringe needle by capillary action; the needle is transported in a sterile dry bottle.

Smears

Best results are obtained if material is spread thinly on clean slides.

Sputum

Send as much sputum (*not* saliva) as possible in a sterile container. If tuberculosis is suspected and sputum is not obtainable consider gastric lavage or a laryngeal swab.

Surgical specimens and biopsies

Send tissue in a dry sterile bottle. (There must be *no* fixative and no disinfectant.)

Urine

Catheter specimens (CSU). Collect urine directly from catheter, not the bag and do not allow the catheter to touch the container.

Midstream specimens (MSSU). To collect a *midstream* urine from females, the patient should be given the following instructions: (1) remove underclothing; (2) wash the vulva and periurethral area with four separate sponges or swabs soaked in 10% soft green soap solution, washing from the front backwards each time (green soap and water wash is preferred to water or saline, and disinfectant solutions must never be used); (3) separate the labia majora with the fingers of one hand and, if necessary, prevent contamination from vaginal secretions by holding a swab between the labia minora; (4) begin to pass urine into the toilet or bed-pan; (5) without interrupting the urine flow, collect the mid-portion of the stream in a wide-mouthed container (wax carton or glass 'honey-pot'); (6) allow the terminal portion of the stream to pass into the toilet or bed-pan. The wide-mouthed container can be sealed and sent directly to the laboratory or urine can be transferred to a sterile disposable universal container.

An MSSU is easily collected from males.

Specimens from infants. Bag specimens are unsatisfactory. The child should be held over a container; an alternative is supra-pubic aspiration. In suspected tuberculosis at least 100 ml of early morning urine (first urine passed) (EMU) on 3 consecutive days should be collected.

Vaginal discharge

Vaginal swabs are taken when pyogenic infection or infection with *Candida* spp is suspected. Sponges should be used only for suspected *Trichomonas* infection. Cervical swabs are taken in uterine infection, salpingitis and pelvic inflammatory disease and to diagnose gonorrhoea, *C. trachomatis* and herpes simplex infections.

INVESTIGATION OF INFECTIVE SYNDROMES

Protocols of investigation

Pyrexia of uncertain origin

This can be defined as a fever, constant or intermittent, with no immediately obvious cause, lasting for at least 4 days. Frequent blood cultures should be taken, 3–4 per day, until the diagnosis is made, and serum samples should be provided for estimations of antibodies to *Brucella, Leptospira, Toxoplasma, Mycoplasma pneumoniae, Coxiella burneti* (Q fever), *Chlamydia* and viruses such as adenovirus, influenza virus, cytomegalovirus, mumps, measles and varicella zoster. Deep sepsis, endocarditis, tuberculosis and exotic protozoal and fungal infections should be considered in the differential diagnosis and the laboratory should be consulted.

Send faeces and urine for culture of *Salmonella*.

Actinomycosis

Analysis of specimens of pus is preferable to processing swabs (a general principle).

Adenopathy, adenitis

Brucellosis and tuberculosis should be considered and serological tests should be performed for infectious mononucleosis and other virus infections as well as for toxoplasmosis. Persistent generalised lymphadenopathy is associated with infection with human T lymphotrophic (HTLV-III) viruses.

Anthrax

In the United Kingdom *malignant pustule* is the commonest presentation. Take swabs from the lesion.

Arthritis

Aspirate joint effusion for culture; paired sera to check for rubella and hepatitis B should be obtained.

Brucellosis

Repeated blood cultures in special medium (*Castaneda*) are the most useful investigations, in addition to serum for agglutinating and complement-fixing antibodies.

Congenital infection

Respiratory secretions, urine, faeces for isolation of bacteria and viruses. Clotted blood for antibody titres to rubella, CMV, herpes simplex virus and *Toxoplasma gondii*. Specimens and clotted blood should be taken from the mother also.

Conjunctivitis

Smears should be made and material for bacterial culture should be plated out at the bedside.

Causative organisms are often non-bacterial, such as *Chlamydia*, herpes simplex virus, adenovirus and molluscum contagiosum virus.

Deep sepsis

Blood culture is useful if the aetiological agent is from inaccessible or unlocated sites.

Diarrhoea

Non-infectious causes should be considered. Infectious causes include *Salmonella*, *Shigella*, *Campylobacter*, enteropathogenic *Escherichia coli* in infants, food-poisoining organisms, vibrios, viruses and amoebae. Repeated stool specimens and blood cultures are required.

Endocarditis

At least 3–4 blood cultures per day for several days are necessary.

If cultures are consistently negative Q fever or psittacosis should be considered and a check should be made to ensure that the patient is not receiving antibiotics. (This should be done in any situation where there are unexpected negative findings).

Food-poisoning

Salmonella spp, for example *typhimurium* and *newport*, *Staphylococcus aureus*, *Clostridium perfringens*, *Bacillus cereus*, *Vibrio parahaemolyticus* or *Campylobacter* spp are the most likely bacterial causes. Stool specimens and any samples of suspected foodstuffs should be sent for examination.

Gangrene

Blood cultures and specimens of pus and necrotic tissues are optimal specimens.

Genital tract infection

Male: urethral, rectal, oral and lesion swabs, scrapes and smears, if indicated for bacteriological and virological examination. Clotted blood for syphilis and viral serology.

Female: Vaginal discharge (swabs and sponge for *Trichomonas vaginalis*) cervical, urethral, rectal and throat swabs, scrapes and smears. Clotted blood for syphilis and viral serology.
N.B. Direct plating for isolation of gonococci on special medium. Use correct transport medium for isolation of viruses and chlamydiae.

Gonoccoccal infection

Smears should be made and cultures taken from all relevant sites; cervix, urethra, rectum, throat and conjunctiva using Thayer-Martin medium or New York City medium. Vaginal swabs are of little use. The Gonococcal complement-fixation test (GCFT) is not reliable.

Hepatitis

Blood is required for hepatitis A antibody (IgM), hepatitis B antigens and antibody tests and paired sera for antibodies to Epstein-Barr virus, herpes simplex virus and cytomegalovirus.

Keratitis

Direct scrapes and swabs, and conjunctival swabs. Remember to use appropriate transport media for virus and chlamydia isolation.

Leptospirosis

Blood culture in special media and serological studies are indicated.

Meningitis

Cerebrospinal fluid (CSF) and blood cultures are needed for bacteriology, and CSF, respiratory secretions, faeces and paired sera for virology.

Osteomyelitis

Blood cultures are essential. Pus should be sent for culture.

Pneumonia

Sputum, for culture of bacterial and fungal pathogens, pharyngeal aspirate, throat washings and paired sera for *Legionella pneumophila*, *Mycoplasma pneumoniae*, psittacosis, Q fever, influenza, adenovirus, RSV, parainfluenza and cytomegaloviruses will cover the range of investigations required for the most likely aetiological agents.

Puerperal pyrexia

Blood, cervical swabs and urine specimens should be cultured.

Rash

Lesion swabs or scrapes or vesicle fluid for bacterial, fungal and viral culture. Respiratory swabs, faeces for isolation and paired sera for viral antibodies.

Septicaemia

Frequent blood cultures are required in addition to screening for underlying causes.

Sore throat

A throat swab and specimen of salvia should be examined for bacteria and viruses; send clotted blood for respiratory virus serology, including a Paul-Bunnell test and specific antibodies to the Epstein-Barr virus and *Toxoplasma gondii*.

Syphilis

10 ml clotted blood are required for screening tests, for example *Treponema pallidum* haemagglutination test (TPHA) and the Venereal Diseases Research Laboratory test (VDRL). Other tests such as the Reiter protein complement fixation test (RPCF) and fluorescent-antibody tests may be available.

Urinary tract infection

Urine (CSU or MSU) should be despatched quickly to laboratory or refrigerated if a delay of more than 4 hours is expected.

Vaginal discharge

Swabs will be examined for *Candida* and pyogenic infection, and exudate (sponge) for Trichomonads.

DIAGNOSTIC APPROACHES IN BACTERIOLOGY

The objects of diagnostic bacteriology are two-fold; firstly, to identify the causative organisms of infection in the patient and, secondly, to establish whether the results of infection, namely antibodies, are present in the patient's serum. Before a specific organism can be accepted as the cause of a specific disease certain criteria, known as the *KOCH-HENLE* postulates, must be satisfied. These state that the organism must be present in the host and be able to be isolated in pure culture from the infected lesions; that the infection can be reproduced when the organisms from the pure culture are inoculated into a susceptible species; and that the organisms can be re-isolated during the course of the experimental disease. These postulates are now supplemented by demonstration of an increase in the antibody titre to the infecting organisms during the course of the disease.

LABORATORY METHODS OF DIAGNOSIS
Direct detection
Inspection

A specimen should be carefully inspected with the naked eye for any features that may be diagnostically significant, such as pus and mucus in the stool of a case of bacillary dysentery and 'sulphur granules' in the pus of a case of actinomycosis. It is also important to ensure that the specimen indicated on the request form is the one that arrives in the laboratory and that it has been properly taken; that a specimen of sputum is not a specimen of saliva, that a specimen of tissue is not in fixative and that a rectal swab does not appear lily-white.

Microscopy

Direct wet films are important in the examination of faeces for cysts and parasites and vaginal exudates for trichomonads. Fixed films are routinely made in the examination of sputum, urine and wound swabs and are commonly stained by Gram's method; they reveal the presence of pus cells, red blood cells and organisms. Other stains can be used such as the Ziehl-Neelsen for tubercle bacilli, Albert's for corynebacteria and methyl violet for Vincent's organisms. In the examination of skin lesions, faeces and throat swabs the Gram film may not be helpful although it can indicate the appropriate method of culture. Only very occasionally can it be reasonably diagnostic, as in the demonstration of Vincent's organisms from a mouth swab or gingival scrapings, Gram-negative intracellular diplococci from the cerebrospinal fluid of a case of meningitis and *Candida*. Histological examination of biopsy sections or exfoliated cells can give an indication of possible viral infection, for example giant cells and nuclear inclusions in lesions of herpes simplex, varicella zoster and CMV viruses.

Phase contrast microscopy is useful in the investigation of spirochaetal infections and *fluorescent antibody techniques* are of value in the direct demonstration of organisms in clinical material, namely *Bordetella pertussis* and respiratory viruses in nasopharyngeal secretions, *Treponema pallidum* in exudates and *Neisseria gonorrhoeae* in urethral discharge. *Electronmicroscopy* is valuable in the diagnosis of (a) viral infection of the skin, for example herpes simplex, varicella zoster, molluscum contagiosum and orf and (b) viral gastroenteritis after concentration of a stool extract by ultracentrifugation.

Detection of antigens and nucleic acids in tissues

The antigens of some bacteria and viruses can be detected directly in the specimen by a variety of immunological tests. Counter-immunoelectrophoresis has been used to detect *Haemophilus influenzae* in CSF and *Streptococcus pneumoniae* antigens in blood. However, much more extensive use has been made of this approach in virological diagnosis. Tests are available using immunofluorescence for respiratory virus infection, and for chlamydial infection, while radio- and enzyme-immunoassays have been developed for the diagnosis of hepatitis B in a serum sample and rotavirus in stool samples. The RNA of rotavirus can be detected after electrophoresis in stool extracts and the DNA of hepatitis B virus in the serum of cases and carriers. The technique may be extended to other viruses which are difficult or slow to grow in culture.

Culture and identification

Culture is important in the diagnostic process and the specimen may have to be processed before culture; for example, sputum is homogenised, many fluids are centrifuged and the deposit cultured. Blood agar is a commonly used medium in medical bacteriology, but many others are available, including *selective media*; these incorporate substances that prevent the growth of certain species but allow the growth of others. The addition of crystal violet to blood agar may be useful in the diagnosis of skin infection as it inhibits staphylococci but permits the growth of *Streptococcus pyogenes*. Similarly, the addition of potassium tellurite makes the isolation of corynebacteria from a throat swab more possible by inhibiting growth of many of the throat commensals. Nalidixic acid and sodium azide can be added to culture media to inhibit Gram-negative bacilli, and the incorporation of certain aminoglycosides such as gentamicin and neomycin in culture media facilitates growth of clostridia and beta-haemolytic streptococci by inhibiting growth of commensal flora.

Blood agar plates may be incubated aerobically or anaerobically and frequently both methods are used. Culture plates can be incubated anaerobically in an anaerobic cabinet or in anaerobic jar from which oxygen has been removed and replaced either by hydrogen or a hydrogen-carbon dioxide mixture from a gas cylinder. Commercially-produced envelopes are available which produce anaerobic conditions within the jars when water is added. Liquid media such as Robertson's meat broth also produce anaerobic conditions in the meat particles at the bottom of the tubes. Finally, many cultures grow better in an atmosphere of 5–10% carbon dioxide, for example gonococci, and many laboratories now use incubation in carbon dioxide for routine specimens.

Identification of isolates

Biochemical tests

Many bacteria have a characteristic biochemical profile which may be value in their identification. A great many of these are known, including:

Enzyme production

A few examples are coagulase from *Staphylococcus aureus*, lecithinase from *Clostridium perfringens* and oxidase from *Neisseria gonorrhoeae*.

Sugar fermentation tests

These are useful in the identification of the enterobacteriaceae. Various sugars are used and are checked for the production of acid, gas or both by the bacteria.

Identification of metabolic products

This can be done by various forms of chromatography, such as gas liquid (GLC) and high pressure liquid (HPLC).

Detection of antigen

Agglutination tests on slides and in tubes and precipitation methods in capillary tubes and agar gel are available. Living suspensions of the unknown organisms are used and these are tested against a range of antisera of known identity. Fluorescent antibody and coagglutination tests for certain antigens may also be performed.

In recent years *counterimmunoelectrophoresis* (*CIE*) radioimmunoassay (RIA), and enzyme-linked immunosorbent assay (ELISA) have been used in the detection of many antigens, such as those present in meningococci, pneumococci, streptococci, salmonellae, *Haemophilus influenzae* and *Proteus* spp.

Special tests applied to isolates

These include epidemiological markers such as bacteriophage typing for staphylococci, colicine production by *Shigella sonnei* and pyocine production by *Pseudomonas aeruginosa*.

Animal tests

In bacteriology these are used to test whether organisms such as clostridia and corynebacteria are toxigenic and in the diagnosis of tuberculosis. They may be useful in virology, for example in the isolation of Coxsackie group A viruses by intraperitoneal inoculation of suckling mice.

Antibiograms

The laboratory identification of organisms is incomplete without determining their antibiotic sensitivity patterns. Different strains of the same bacterial species may be identified by their different antibiograms.

The culture of viruses and chlamydiae

Specimens for virus isolation are treated with antibiotics and an antifungal agent to prevent bacterial and fungal contamination of the cell cultures. Several cell cultures are inoculated; for example human embryo fibroblasts, monkey kidney and human epithelial cells are widely used to isolate the common viruses. Cells are usually grown as monolayers in tubes at 37°C although respiratory viruses grow best at 32°C. Growth of the virus is detected by microscopy for a cytopathic effect (CPE). The damage to the cells causes rounding and eventually death and detachment of the cells from the tube. The cells may be swollen, round, irregular or shrunken and some viruses fuse cells to form syncytia. The appearance of the CPE, its progression and the cell type give an indication of the identity. A few can be detected within 24–48 hours, but many will take up to 7–10 days. Some, such as cytomegalovirus, take longer. Some viruses do not produce a CPE, but may be detected by haemadsorption, interference, immunological or biochemical tests. Animal inoculation is seldom done, but can be used for the isolation of Coxsackie A and togaviruses. Similarly the use of embryonated eggs has been replaced by cell culture. It is important to remember to take specimens from patients for virus isolation as early as possible in the illness as virus excretion often declines rapidly over a few days after the onset of symptoms.

Detection of antibody

The detection of rising titres of antibody specific to a microorganism can be used to confirm a current or recent infection with an organism which may have been grown from the patient. This approach is used to diagnose infection with an organism which cannot be grown or which is dangerous to grow. If the patient presents soon after the onset of symptoms it is essential to submit an acute sample and one collected later—up to 10–14 days. The exact timing depends on the nature of the infection: thus in infections with short incubation periods (2–3 days) the antibody response will not be detectable for a fortnight. However with most systemic infections such as rubella and chickenpox, the antibody may even be present when the patient presents and the titre will rise rapidly over the next week. In a few diseases, for example hepatitis A, with an even longer incubation period of about 3–4 weeks, the antibody response has already occurred and it is unusual to demonstrate a rising titre. In such cases, or when only a single specimen is available, the detection of specific IgM antibody to the virus is of diagnostic significance. Such tests are of value in the diagnosis of congenital and neonatal infection and in rubella, cytomegalovirus and hepatitis A and B virus infections and toxoplasmosis.

In some infections such as Q fever endocarditis the titres may be so high as to be diagnostic. Interpretation of serological tests is entirely dependent on knowledge of the illness and its duration when the serum specimens were collected.

A wide range of immunological tests have been applied including simple agglutination tests with bacterial suspensions and extracts and complement fixation for many viruses, mycoplasma, Q fever and brucellosis, toxoplasmosis and syphilis.

Immunofluorescence has been widely used for detection of viral antibodies, for example to Epstein-Barr virus and other herpes viruses. Other tests that have proved of value for detection of antibody in more recent times are RIA and ELISA. These are sensitive and have been developed for antibody and IgM antibody detection. They have been of great value in the diagnosis of hepatitis A and the antibody responses to various hepatitis B antigens.

Non-specific tests are also available, including the *Paul-Bunnell* test for infectious mononucleosis and the *Wassermann Reaction* for syphilis. Tests can also be done on the patient, for example the *Schick test* which demonstrates antitoxic immunity to diphtheria, and the *tuberculin test* which tests for previous contact with *Mycobacterium tuberculosis*.

Interpretation of laboratory findings

Interpretation of findings is not always an easy matter. In some instances it can be relatively simple as, for example, the isolation of *Staphylococcus aureus* from a septic skin lesion, *Mycobacterium tuberculosis* from the sputum of a patient with respiratory symptoms or bacteria and viruses from CSF. On the other hand the isolation of *Neisseria meningitidis* from the throat of a patient with meningitis, or *Streptococcus pyogenes* from the throat of a patient complaining of sore throat may merely indicate simple carriage of these organisms; viruses may have been the causes of infection in both cases. Detection of hepatitis B surface antigen in the blood indicates that the patient is suffering from an acute infection or that he is a long-term carrier. Further tests or follow-up are needed to clarify this situation. A single isolation of *Staphylococcus aureus* from the blood of a patient with suspected infective endocarditis may not be of diagnostic significance but repeated isolations of the organisms of the same phage type confirm the diagnosis. Knowledge of the normal flora of the various areas of the body can frequently resolve the problem of identifying which organisms are contaminants, and information as to the approximate numbers of organisms in a specimen is helpful because contaminants are frequently present in smaller numbers than the organism causing the disease. Enteroviruses may be excreted for some weeks after an unrecognised infection. Caution is essential before concluding that such an isolation indicates the cause of a current infection such as meningitis. Antibody tests can also be used for clarification of the cause. The response to antibiotics chosen on the basis of antibiograms of the organisms presumed to be the cause may be helpful. For example, failure of response to penicillin in a patient with pneumonia and penicillin-sensitive pneumococci should raise the suspicion that some other agent such as penicillin-resistant *Staphylococcus aureus* may be responsible.

Increasingly the role of endogenous organisms in infections is being realised and this makes valid interpretation of laboratory findings difficult. In this complex area close clinical and laboratory collaboration is essential.

FURTHER READING

Duguid J P, Marmion B P, Swain R H A 1978 Medical microbiology, 13th edn. Churchill Livingstone, Edinburgh

Passmore R, Robson J S 1973 A companion to medical studies. Blackwell Scientific Publications, Oxford

Rytel M W 1975 Rapid diagnostic methods in infectious diseases. Advances in Internal Medicine 20: 37–60

Stokes E J 1975 Clinical microbiology. Arnold, London

5

Treatment of infections

ANTIBIOTICS AND CHEMOTHERAPY

Principles of use

Is an antibiotic necessary?

Antibiotics are often administered totally unnecessarily. In the author's experience they have been given for the common cold, flea bites, sebaceous cysts, a sprained ankle and bleeding from the urinary tract. Antibiotics, frequently prescribed for such lesions as a superficial collection of pus, cannot penetrate formed pus, nor can they penetrate adequately fibrin and granulation tissue. Usually simple surgical drainage is all that is required. In deciding whether or not to use an antibiotic toxicity reactions and side effects must be kept in mind and although some antimicrobial agents produce more serious side effects than others none is absolutely safe. Even penicillin which is accepted as one of the safer antibiotics can produce serious and sometimes fatal hypersensitivity reactions in a few people. The advantages and disadvantages of antibiotic therapy must be weighed up carefully for each individual.

The choice of drug

This depends primarily on the nature of the infecting organism and its antibiotic sensitivity pattern. Antibiotics should be used on a rational bacteriological basis although there is a place for an *educated guess* in a clinical emergency such as septicaemia, or osteomyelitis in a child. Consideration must be given as to whether a bactericidal or bacteriostatic drug of broad or narrow spectrum should be used.

Bactericidal and bacteriostatic action. Antimicrobial drugs may be bactericidal and rapidly kill bacteria, or bacteriostatic and inhibit multiplication without actually killing the organisms. The line of demarcation between the two actions is not clear cut and is dependent on factors such as the extent of the bacterial challenge, the concentration of the drug and the organism concerned; some drugs such as erythromycin are bacteriostatic at normal dosage but are bactericidal in higher concentrations. Theoretically bactericidal drugs are to be preferred to bacteriostatic drugs; however, a critical factor in the cure of infection is the patient's own defence system and bacteriostatic drugs may be able to inhibit multiplication of the bacteria sufficiently to allow their elimination by the white cells.

Broad-spectrum and narrow-spectrum drugs. Antimicrobial drugs differ in the range of bacteria they inhibit. Broad-spectrum drugs such as the tetracyclines and ampicillin have a wide range of activity against both Gram-positive and Gram-negative bacteria, whereas flucloxacillin has a narrow spectrum. It is very effective against *Staphylococcus aureus* but ineffective against almost all Gram-negative bacteria. Narrow-spectrum drugs are to be preferred because of their greater specificity and unlike broad-spectrum drugs they do not have the same adverse effects on the normal flora of the body. Broad-spectrum drugs can so alter the composition of the flora at various sites that its normal protective effect is ruined, and it is well recognised that certain pathogenic organisms can colonise such areas and produce superinfection, as in infection of the mouth by *Candida* spp.

Combinations of drugs

In general a single antimicrobial drug should be used for the treatment of a particular infection but there are indications for the use of a combination of drugs. These are: (1) to achieve a synergistic effect, the combination of drugs producing a much greater effect than either drug alone; (2) to prevent the development of bacterial resistance; (3) to deal with mixed infections.

In the treatment of infective endocarditis caused by *Streptococcus faecalis* it has been shown that a combination of a penicillin and an aminoglycoside is often therapeutic whereas this would not be so with full doses of either drug alone. Synergism is well demonstrated in the laboratory in the combination of sulphonamide and trimethoprim in the drug *cotrimoxazole*. The use of combinations of drugs to prevent the emergence of drug-resistant strains is well demonstrated in the treatment of tuberculosis. In this disease treatment is prolonged and resistance to single drugs occurs quickly.

The incidence of mutants resistant to a single drug may be in the order of $1:10^7$ cell divisions so that if two drugs are given the chances are then only $1:10^{14}$.

Combinations of drugs should not be used either empirically in mixed infections or when the diagnosis is obscure, although an important exception is when a person is seriously ill and no bacteria have been isolated. Some combinations theoretically may be *antagonistic*; for example, penicillin acts on organisms that are actively growing and if it is combined with a bacteriostatic drug this action may be prevented.

Fixed-dosage preparations (commercial preparations) must be used with caution, because the proportion of each drug may be wrong for the particular infection to be treated and there is no scope for altering these proportions.

General considerations

Antibiotics should be given in therapeutic doses, for an adequate period. If after a few days treatment there is no clinical response it usually means either that the infecting organisms are resistant, in which case further specimens should be sent to the laboratory, or that the drug is failing to reach the bacteria in sufficient concentration. *Local treatment* should be carried out with antibiotics such as neomycin that are not used systemically, because if drug resistance occurs a valuable systemic drug becomes useless for treatment of serious infection in that patient. *Side effects* of drugs should be known and the drug of choice should be the one that incorporates minimal side effects with maximum efficiency, although this is not always possible with organisms that are multiply-resistant. *Drug incompatibility* must be remembered. Nalidixic acid, for example, potentiates the action of warfarin and severe bleeding may ensue. The combination of alcohol and metronidazole has an antabuse effect.

Prophylactic use of drugs

Surveys conducted on aspects of prophylaxis have frequently reported that an increase in resistance to the antibiotics used can occur and that widespread prophylactic use may predispose to infections with more resistant organisms by altering the composition of the protective commensal flora or by selecting out resistant mutants. Ironically a substantial number of reports have stated that patients on prophylactic drugs, particularly those undergoing surgery, have a higher infection rate than those in non-treated control groups because the antibiotics prescribed are frequently totally inappropriate for the bacteria that subsequently cause the infections.

There are however some clear and justifiable indications for prophylaxis. In trauma, when there are the possibilities of tetanus or gas gangrene if the wound becomes contaminated by *Clostridium tetani* or *Clostridium perfringens*, benzyl or phenoxymethyl penicillin should be given. Following thigh amputation in the elderly there is a risk of endogenous infection from *Clostridium perfringens* and a similar policy should be followed. Children who have had a primary attack of rheumatic fever run a high risk of a subsequent attack and phenoxymethyl penicillin should be given to them to prevent reinfection with *Streptococcus pyogenes*. Antibiotics have proved useful in the prevention of infective endocarditis in those with damaged valves, valve prostheses or congenital malformations, when undergoing dental extraction.

In young girls and women who are prone to urinary tract infection antibiotic prophylaxis may prevent recurrent attacks. There is now evidence that antibiotic cover in large bowel surgery is useful and in this regard metronidazole is given to deal with the anaerobes; however, the choice of antibiotic to be effective against the aerobic organisms must still remain a matter of guesswork.

Within these widely different sets of circumstances there are grey areas where there can be arguments for or against prophylaxis, as in open heart surgery and before anaesthesia in patients with chronic respiratory disease. It is desirable to continue chemotherapy for Q-fever endocarditis during the operation and thereafter for many months.

It should be recognised that the inherent dangers of the prophylactic use of antibiotics may far outweigh the mental balm and sense of security produced by the use of the prophylactic agent.

Harmful effects of antibiotics

These include the production and selection of antibiotic resistant bacteria, which are then spread by cross infection, particularly in hospital, toxicity reactions and the development of hypersensitivity.

Mode of action of antimicrobial agents

There are five main modes of action of antibiotics:

Inhibition of cell-wall synthesis

Drugs that prevent cell-wall synthesis do so by inhibiting the transpeptidase step required for cross-linking the polysaccharide chains in the peptidoglycan layer of the cell wall. The *penicillins; cephalosporins* and *vancomycin* act in this way.

Damage to cell membrane

Some antibiotics produce damage by binding to the membrane, acting as cationic detergents, resulting in the loss of the semipermeable properties of the membrane. Substances then pass out from the cell and death occurs. Polypeptide antibiotics such as *polymyxin* act in this fashion.

Inhibition of protein synthesis

Antibiotics causing this have an effect on *translation* in the cell and include *chloramphenicol*; the *tetracyclines*; the *aminoglycosides* such as *streptomycin, neomycin, kanamycin, amikacin, tobramycin, metilmicin* and *gentamicin; erythromycin; lincomycin* and *clindamycin*.

Inhibition of nucleic acid synthesis

Nalidixic acid inhibits replication of DNA and rifampicin blocks transcription in prokaryotic cells.

Inhibition of folate synthesis

Sulphonamides and *trimethoprim* block different steps in folate synthesis in bacteria. Folic acid is needed as a cofactor in the synthesis of thymidine and other nucleotides.

Resistance of bacteria to antimicrobial drugs

Bacteria may be *naturally* resistant to a particular antibiotic but they may also *acquire* resistance to drugs. The two mechanisms whereby bacteria initially sensitive to an antibiotic may become resistant are by *mutation* and *gene transfer*.

Mutation

Mutations develop spontaneously in bacterial cultures at a

frequency of one per 10^7–10^9 cell divisions, whether or not a drug has been administered. The resistant mutants that occur in the presence of antibiotic are not as robust as the parent strains and do not survive for long if the drug is discontinued. If, however, there is continued exposure to the antibiotic, sensitive parent strains will be killed whereas the resistant mutants will remain viable and multiply. The action of the drug therefore is to *select* the resistant bacteria.

The mechanism of resistance depends on the mode of action of the drug. For example, mutations may produce decreased affinity of the target site reduced penetration of the drug into the cell or destruction of the drug by enzymes.

Gene transfer

This can occur by *transduction* or *conjugation* and consists of transfer of genetic material from a strain or species previously resistant to one or more antibiotic to a strain previously sensitive. In these cases genes for drug resistance are carried on extrachromosomal elements or *plasmids*.

Transduction

Penicillin resistance in *Staphylococcus aureus* depends on the presence of a plasmid carrying the gene for penicillinase (beta-lactamase) production. These plasmids can be transferred amongst different staphylococcal strains by bacteriophages.

Conjugation

R-factors (plasmids bearing drug-resistance genes) may be spread by conjugation amongst cells of a wide range of species, notably the enterobacteria and *Pseudomonas aeruginosa*. Contact between donor and recipient cells is required and the genetic material is transferred directly between the two cells.

Selection

Selection by antibiotics is important whether resistance is produced by mutation or gene transfer. Repeated exposure of bacteria to the drug together with long periods of exposure in the population means that more resistant strains will be selected out. An important step in limiting the increasing number of resistant strains is to reduce the use of the drug. This should allow the return of the sensitive strains.

Laboratory control of antibiotic therapy

Such assays include tests on the organism as well as measurement of responses in the patient.

Tests on the organism

These are essentially of two types, *dilution tests* and *disk sensitivity tests*.

Tube dilution tests. These methods are too time-consuming for routine use but they are carried out from time to time in all laboratories. They indicate the minimum inhibitory concentration (MIC) of a drug required for a specific organism in various diseases, such as infective endocarditis, thereby ensuring a more accurate and effective dose regime. The MIC is read from the tube containing the highest dilution of the antibiotic to inhibit growth.

Agar plate dilution tests. These are useful when a large number of strains have to be tested. Various dilutions of the antibiotic are prepared in agar plates and each plate is seeded with several strains. After incubation the MIC of the drug can be readily ascertained from the plate that contains the highest dilution of the antibiotic required to prevent growth of the test strain.

Disk diffusion tests. These are inherently less accurate than dilution tests but are more rapid and convenient. Disks are made of filter paper impregnated with a known amount of antibiotic and placed on an agar plate that has been inoculated with the test organism. After incubation the plate is examined for zones of inhibition around the disks where the antibiotic has diffused into the medium. Standardisation of results is important in such tests and this can be done in one of two ways: (1) correlation of the diameter of the zone with a scale relating zone size to the MIC. Examples are the Ericsson and Kirby-Bauer methods; (2) the zone size produced by the organism is compared with the zone size given by a control organism tested by the same method. Many variables can affect the results of such tests, such as inoculum size, composition of the medium, antibiotic content of the disk and pH, but the main advantage of such tests is that the clinician can be informed of the results in the shortest possible time.

TREATMENT OF INFECTIONS   81

Tests of responses in the patient

Assays can be done on the serum, urine, sputum or cerebrospinal fluid and these measure whether effective concentrations of an antibiotic are being attained. They are also useful in guarding against excessively high serum levels of antibiotics with serious side effects. *Tube dilutions* and *agar diffusion* methods may be used.

In *tube dilutions*, graded dilutions of the fluid to be tested, usually serum, are made in a liquid medium; a standard inoculum of bacteria of known sensitivity, usually NCTC strains, for example *Staphylococcus aureus*, NCTC 6571, the *Oxford* staphylococcus, is then added. After incubation the tubes are examined for turbidity and the tube that contains the highest dilution of serum to inhibit growth of the organisms is noted. The MIC of the Oxford staphylococcus is known for each antibiotic and this allows calculation of the amount of antibiotic present in the patient's undiluted serum. It can also be helpful to check the patient's serum levels of antibiotic against the organism isolated.

Agar diffusion methods. In these holes are cut in the agar previously inoculated with a control organism. The holes are filled with the serum under test and zones of inhibition produced by the antibiotic in the serum are compared with a graph of the sizes of zones produced by known concentrations of the antibiotic.

In guarding against excessive serum levels of a potentially toxic antibiotic, estimations can be made by testing serum taken about one hour after a dose (*peak level*) and serum taken just before the next dose (*trough level*). *Rapid assays* may from time to time be required particularly in monitoring treatment with aminoglycosides during renal failure when results may be needed on the same day. Rapid plate assays are available and other methods include a *urease assay, radio-immune assays, radio-tagging assays* and an *immunofluorescence polarisation assay*.

ANTIMICROBIAL DRUGS OF CLINICAL IMPORTANCE

Sulphonamides

Antibacterial spectrum and uses

These drugs have a broad-spectrum, bacteriostatic effect. They are effective against certain Gram-positive organisms such as pneumococci but their use against these organisms has been superseded by other drugs. Some Gram-negative cocci are sensitive and likewise many enterobacteria. Sulphonamides are used against Gram-negative bacilli, principally in urinary tract infection caused by *Escherichia coli*.

Mode of action

Sulphonamides are structural analogues of para-aminobenzoic acid (PABA) which is involved in the synthesis of folic acid by bacteria. They inhibit bacterial growth by competitive inhibition of the enzymes involved in the incorporation of PABA into the folate pathway. Humans cannot synthesise folates and have to absorb them from the intestine, thus sulphonamides show selective toxicity by inhibiting a synthetic pathway not present in the host's cells.

Resistance

Many species, particularly meningococci, in the past sensitive to the sulphonamides, are now resistant and this applies also to other bacteria. Resistance can be produced by synthesis of a folic acid synthetase that has a lowered affinity for sulphonamide, or by overproduction of PABA. Cross resistance in the group is complete.

Cotrimoxazole

Antibacterial spectrum and uses

Cotrimoxazole is a synthetic oral and parenteral antimicrobial agent and is a combination of a *sulphonamide* (sulphamethoxazole) and *trimethoprim* and has a bactericidal action. Although it is effective against certain Gram-positive species and *Haemophilus influenzae* cotrimoxazole is particularly useful against many of the enterobacteria that cause *urinary tract infection* although many hospital coliforms and anaerobes are now resistant. It can also be used in typhoid fever and brucellosis.

Trimethoprim is now being used increasingly for respiratory and urinary tract infections because it has fewer side effects than cotrimoxazole.

Mode of action

Trimethoprim is a synthetic agent that in concert with sulphonamides produces a sequential blockade of folic acid synthesis in bacteria. The combination produces a strongly-synergistic action and the effect of the combination is equal to about ten times the effect of trimethoprim alone.

Resistance

Although this can be produced easily in the laboratory by the transmission of R-factors it is not yet a problem clinically.

The penicillins

The following will be considered: *benzyl penicillin, phenoxymethyl penicillin,* the *cloxacillins,* the *ampicillins* and the *carbenicillins.*

Mode of action (all penicillins)

Penicillins act by inhibiting the transpeptidase step required for cross-linking the polysaccharide chains in the peptidoglycan layer of the cell wall. All are bactericidal.

Benzyl penicillin (penicillin G)

This was the original penicillin and is still the most effective against *Streptococcus pyogenes* and non-penicillinase producing *Staphylococcus aureus,* pneumococci, clostridia and many neisseriae. It is useful against *Bacillus anthracis* and *Actinomyces israelii* but Gram-negative bacilli such as the enterobacteria and *Pseudomonas* spp are highly resistant to it. It is inactivated by gastric acid and has to be given parenterally.

Resistance. Many strains of *Staphylococcus aureus* are now resistant because of the production of a penicillin-destroying enzyme (beta-lactamase) that hydrolyses the beta-lactam ring in the penicillin nucleus. Such strains are much more common in hospital and are spread by cross infection and selected out by continued use of the drug. In the community there is increasing resistance to penicillin in gonococci due either to mutation or the production of a beta-lactamase obtained via R-factors from *Escherichia coli.* Some Gram-negative bacilli produce beta-lactamases and this makes them resistant to benzyl penicillin and also to other penicillins such as ampicillin.

Phenoxymethyl penicillin (penicillin V)

Antibacterial spectrum and uses. This drug can be given orally because it is acid stable and it has much the same antibacterial spectrum as benzyl penicillin, although in serious systemic infection it is preferable to give high doses of benzyl penicillin because absorption of penicillin V from the gut is unpredictable. Phenoxyethyl and phenoxypropyl penicillins exist but none is better than phenoxymethyl.

Resistance. This is the same as for benzyl penicillin.

The cloxacillins

Antibacterial spectrum and uses. These belong to the group of semi-synthetic penicillins that are produced by chemical manipulation of the penicillin nucleus. *Methicillin* was the first of this group to be introduced, then *cloxacillin* and *flucloxacillin*. Methicillin has now been largely superseded as it has to be given by injection because of its instability in gastric juice. Cloxacillin can be given orally or parenterally; flucloxacillin is a pharmacological improvement on cloxacillin. These penicillins are used against penicillinase-producing staphylococci because their penicillin molecule is not affected by penicillinase. For infections caused by penicillin-sensitive organisms they are less effective than benzyl or phenoxymethyl penicillins.

Resistance. Resistance, though uncommon, does occur clinically; this is not due to production of a beta-lactamase, but probably to mutation.

The ampicillins

Antibacterial spectrum and uses. This semi-synthetic penicillin has a far wider spectrum of activity than the other pencillins and is a *broad-spectrum antibiotic*. It is destroyed by penicillinase and is therefore of no value against the penicillinase-producing *Staphylococcus aureus*. In general, it is much less active than benzyl penicillin against most Gram-positive cocci and should not be used for infections with such organisms, although an exception is infection caused by *Streptococcus faecalis*. It can be used against many Gram-negative organisms such as *Escherichia coli* (although resistance is increasing), *Haemophilus influenzae*, non-penicillinase-

producing *Proteus* spp *Shigella* and *Salmonella*. It is not absorbed well from the gut.
Resistance. Many enterobacteria are now resistant due to R-factor beta-lactamase production.

Amoxycillin

Antibacterial activity is similar to ampicillin but the absorption of this drug is considerably greater than ampicillin and is not influenced by the presence of food.

Augmentin

This is a combination of amoxycillin and clavulanic acid; the latter inactivates penicillinases, and augmentin can therefore be used orally against a broad spectrum of beta-lactamase producing bacteria.

Talampicillin

This oral drug has an antibacterial activity similar to ampicillin but, like amoxycillin, is much better absorbed from the gut.

Mezlocillin and azlocillin

Both are substituted penicillins given intravenously, and have a good broad spectrum activity. They are particularly effective against *Pseudomonas* spp.

Piperacillin

This parenteral antibiotic is related to mezlocillin and azlocillin and has a strong activity against *Pseudomonas* spp.

Carbenicillin

Antibacterial spectrum and uses. This semi-synthetic penicillin has a special use in *Proteus* and *Pseudomonas* infections in particular, but is ineffective against Gram-positive organisms. It may be given parenterally.
Resistance. Carbenicillin is inactivated by penicillinase.

Carfecillin

This is a phenyl ester of carbenicillin and can be given orally. It is rapidly excreted in the urine and is therefore not suitable as a systemic antibiotic; it is however useful in the treatment of infections of the lower urinary tract.

Ticarcillin

This derivative is active against *Proteus* and *Pseudomonas* organisms and is twice as active as carbenicillin against susceptible organisms.

Other penicillins

Mecillinam is active against Gram-bacilli except *Pseudomonas* and is effective in treating intestinal infections such as typhoid, and urinary tract infections. It is given parenterally. *Pivmecillinam* is an oral analogue.

The cephalosporins

Antibacterial spectrum and uses

The cephalosporin family of antibiotics is related to the penicillins. Many semi-synthetic derivatives are available with different properties, but the available compounds do not show such marked differences in their antibacterial spectrum as in the penicillin family. *Cephaloridine* and *cephalothin* are given parenterally and *cephalexin* orally. All have a broad spectrum of activity inhibiting both Gram-positive cocci and Gram-negative bacilli, but most infections are more effectively treated with the appropriate penicillin unless the patient is hypersensitive to penicillin. This would seem to be the main indication for using cephalosporins, although cross hypersensitivity may occur.

There are so many cephalosporin drugs on the market that the choice of any one is difficult. The original cephalosporins (the 'first generation') were *cephaloridine* and *cephalothin*, both of which are administered parenterally, and *cephalexin* given orally. An improvement on these was brought about by the introduction of 'second generation' cephalosporins that were less susceptible to the effect of beta-lactamase producing organisms. These include *cefamandole*, *cefuroxime* and *cefoxitin* (a member of a group of beta-lactam antibiotics the *cephamycins*, that resemble the cephalosporins). All these are given parenterally.

In recent years a 'third generation' has been developed, including *cefotaxime, ceftazidime* and *ceftizoxime*, drugs given parenterally. This generation has a wider spectrum of activity than earlier drugs and they act against enterobacteriaceae (including *Pseudomonas*) and anaerobes such as *Bacteroides*. They are also strongly effective against beta-lactamase producing organisms and are less toxic.

Mode of action

Cephalosporins are bactericidal, acting on the growing cell wall in a similar manner to the penicillins.

Resistance

Resistance in some Gram-negative species is intrinsic but there is also some enzyme destruction of the antibiotic by the production of a beta-lactamase (cephalosporinase) by some Gram-negative bacilli, although the cephalosporins tend to be less susceptible to the beta-lactamase produced by staphylococci.

The aminoglycosides

These include *streptomycin, neomycin, framycetin, kanamycin, amikacin, gentamicin, tobramycin, sissomicin* and *netilmicin*. They are active against many Gram-negative bacilli and staphylococci but not against streptococci or against anaerobic bacteria. Except for neomycin and framycetin all are administered parenterally. Important side effects are nephro- and ototoxicity.

Streptomycin

Antibacterial spectrum and uses. This drug has a high degree of activity against Gram-negative bacilli, but the newer aminoglycosides such as kanamycin and gentamicin are now used more frequently for such infections. Streptomycin is still important in the treatment of tuberculosis and can be used also in synergistic combination with penicillin or ampicillin in the treatment of infective endocarditis caused by *Streptococcus faecalis* and the viridans streptococci, although combinations of other penicillins and aminoglycosides have been recommended.

Neomycin and framycetin

Antibacterial spectrum and uses. These antibiotics are similar in most respects. They are too toxic for systemic use and are therefore applied topically; both can be used to suppress nasal carriage of *Staphylococcus aureus* and they may be useful for superficial infections caused by coliform bacilli. Neomycin, because it is poorly absorbed from the gut, can be given before intestinal surgery in an attempt to suppress growth of organisms in the intestine. Whether attempted pre-operative 'sterilisation' of the intestine with this drug is of any benefit is open to serious doubt.

Kanamycin

Antibacterial spectrum and uses. As with neomycin it can be used to suppress the aerobic Gram-negative bowel flora preoperatively. It is effective in treatment of severe systemic infections by Gram-negative bacilli, particularly those caused by *Proteus* spp, and is of considerable use in the treatment of those severe urinary tract infections that may be associated with septicaemia.

Amikacin

This is a derivative of kanamycin but with a slightly broader spectrum of action, in that *Pseudomonas* spp are sensitive to it. It is a most useful drug in hospital practice in the treatment of many Gram-negative infections resistant to gentamicin.

Gentamicin

Antibacterial spectrum and uses. This drug has a broader spectrum of activity than the other aminoglycosides and a lower inhibitory concentration is required against *Escherichia*, *Klebsiella*, *Proteus* and *Salmonella* spp; it is more effective against *Pseudomonas* spp and *Staphylococcus aureus*. In serious pseudomonas infections it may be given in combination with carbenicillin. It is also given in combination with penicillin or ampicillin in the treatment of infective endocarditis caused by viridans streptococci or *Streptococcus faecalis*.

Tobramycin

Antibacterial spectrum and uses. This recently introduced antibiotic may prove more effective than gentamicin in the treatment of pseudomonas infection.

Sissomicin

This is similar to gentamicin and does not appear to be more effective.

Netilmicin

A derivative of sissomicin, this drug is reported to be effective against many gentamicin-resistant strains and is less toxic than gentamicin.

Spectinomycin

This is related to the aminoglycosides and its main use is in the treatment of gonorrhoea.

Mode of action. All aminoglycosides are bactericidal and act by inhibiting protein synthesis. They have some serious toxic effects on the eighth cranial nerve and kidney related to the blood levels.

Resistance. Mutation produces ribosomes that are no longer inhibited by the drug. R-factor resistance, by production of enzymes that inactivate aminoglycosides, occurs in enterobacteria. There can be cross resistance but this is not complete within the group.

The tetracyclines

These comprise a group of closely-related oral antibiotics that include *tetracycline, chlortetracycline, oxytetracycline, doxycycline, minocycline* and *demethylchlortetracycline*. Their antibacterial activity is very similar and there is little to choose amongst them.

Antibacterial spectrum and uses. Their spectrum of activity is broad. They are effective against Gram-positive and Gram-negative bacteria, both aerobes and anaerobes, as well as against mycoplasmas, chlamydiae and *Coxiella burneti*, although many *Bacteroides* spp are now resistant. They are commonly used in acute exacerbations of chronic bronchitis because of their activity against *Haemophilus*.

Mode of action. The tetracylines are bacteriostatic and inhibit protein synthesis by prokaryotic ribosomes.

Resistance. Acquired resistance is common now in bacteria that were previously sensitive such as *Staphylococcus aureus*, *Streptococcus pyogenes* and pneumococci.

Chloramphenicol

Antibacterial spectrum and uses. A broad-spectrum antibiotic, chloramphenicol is effective against most pathogenic bacteria, with the exception of *Pseudomonas* spp, but because of its toxic side effects on the bone marrow and its possible role in causing circulatory collapse in infants its use must be severely limited to certain infections, notably *Haemophilus influenzae* meningitis in children and the acute stages of typhoid fever; for both of these infections it is the drug of choice. It can be used for the treatment of coxiella infections and can be given orally.

Mode of action. Chloramphenicol produces its bacteriostatic effect by inhibiting protein synthesis.

Resistance. Resistance to chloramphenicol is present on the R-factors of many coliform bacilli particularly in countries where it is still widely available without prescription.

Erythromycin

Antibacterial spectrum and uses. Like benzyl penicillin this is a narrow-spectrum drug, active primarily against Gram-positive organisms such as beta-haemolytic streptococci, pneumococci, *Staphylococcus aureus* and some Gram-negative anaerobes and clostridia. It is not effective against the enterobacteria and most other Gram-negative bacilli although it is active against neisseriae, *Haemophilus influenzae*, *Bordetella pertussis*, *Legionella pneumophila* and many strains of *Bacteroides* spp; it has also an effect against some chlamydiae, mycoplasmas and rickettsiae. It is particularly useful however in the treatment of patients who are hypersensitive to penicillin. It is given orally or intravenously.

Mode of action. This is mainly bacteriostatic, though high doses may be bactericidal. It acts on bacterial ribosomes to inhibit protein synthesis.

Resistance. In *Staphylococcus aureus*, streptococci and pneumococci this arises rather easily by mutation, though resistant strains usually disappear if the drug is discontinued for a period.

The lincomycins

Lincomycin and *clindamycin* are similar in many ways to *erythromycin* in antibacterial spectrum, mode of action and resistance. Cross resistance occurs amongst these drugs. Clindamycin is the newer, better absorbed and more active drug and has a greater effect against anaerobes. Both are useful in the treatment of infections caused by Gram-positive aerobes, such as *Staphylococcus aureus*, and because of very good penetration and diffusion to tissues such as bone they are recommended for the treatment of acute and chronic osteomyelitis. They are also effective against bacteroides organisms, fusobacteria and anaerobic cocci. A combination of gentamicin and clindamycin may be invaluable in blind treatment of serious undiagnosed infection. Lincomycins have been used in combination with the tetracyclines to treat chronic Q fever infections.

Fusidic acid

Antibacterial spectrum and uses. Fusidic acid is effective against many Gram-positive cocci and Gram-negative bacilli but its outstanding action is against *Staphylococcus aureus*. It has been used successfully in acute and chronic forms of staphylococcal disease and can be given orally or intravenously.

Mode of action. Fusidic acid inhibits protein synthesis.

Resistance. Resistance to the drug is by mutation, and as this can occur quickly fusidic acid should not be given on its own.

The polypeptides

Antibacterial spectrum and uses. These antibiotics are nearly all too toxic for systemic use but can be of value as topical agents in infections of the eye, ear and burns. *Colistin* (polymyxin E), the only member that can be given systemically, is active against *Pseudomonas aeruginosa*. *Bacitracin*, a peptide antibiotic, is too toxic for systemic use but is sometimes included in mixtures of antibiotics for topical use. It is used in the laboratory, incorporated in a paper disk, as a screening test for *Streptococcus pyogenes*.

Rifampicin

This is an effective oral antituberculous drug which should be given in combination with some other drug (for example isoniazid) because resistance to it develops rapidly. It can be used for the treatment of infection by *Legionella pneumophila*.

Vancomycin

Given orally or intravenously, this is a bactericidal drug used against Gram-positive organisms and is very effective in endocarditis caused by Gram-positive cocci, particularly in cases of penicillin hypersensitivity. It is the drug of choice for pseudomembranous colitis.

Novobiocin

This can be used orally or parenterally for infections caused by *Staphylococcus aureus* but should be combined with another antibiotic as resistance emerges rapidly.

Metronidazole

This antimicrobial agent can be given orally, intravenously or by suppository and is very effective against non-sporing anaerobes such as *Bacteroides* and fusobacteria. It can be used also for clostridia and for some anaerobic cocci.

It is bactericidal and acts by interfering with DNA and RNA polymerases. It can be used as a prophylactic agent for surgery of the large bowel, for anaerobic pulmonary infection, abscesses and certain types of post-operative sepsis. It is used also for the treatment of trichomoniasis, giardiasis and amoebiasis.

Nalidixic acid

This agent inhibits DNA replication and acts against coliforms and *Proteus* organisms, but not *Pseudomonas aeruginosa*. It is used solely for urinary tract infections and is given orally.

Nitrofurantoin

This oral agent is also used solely for urinary tract infection. It has an antibacterial spectrum that includes *Streptococcus faecalis* and coliforms but is not active against *Proteus* or *Pseudomonas* spp.

ANTIVIRAL THERAPY

Antiviral therapy is potentially of value (1) in the treatment of severe infections, for example encephalitis, rabies and infections in compromised patients; (2) in chronic diseases, for example carriers of hepatitis B virus, latent HTLV-III (LAV) and latent herpes virus infections of sensory ganglia; (3) to reduce the morbidity of common infections such as the common cold and influenza and (4) to prevent the development of viral or bacterial pneumonias. The need to commence therapy as soon as possible is an important general consideration. By the onset of symptoms, in both local and disseminated infections, extensive virus replication and host cell damage has occurred and it will be difficult to effect a dramatic reduction in the time of recovery. In some situations, very early treatment may abort infection, for example the recrudescence of cold sores. Prophylactic use is likely to be a more realistic objective in respiratory tract infections.

Worthwhile benefits of therapy could arise in treating patients who cannot control infections which therefore run a protracted course. Herpetic infections in immunocompromised patients are good examples of this problem.

There are few acceptable antiviral agents available due to the difficulty in finding drugs that will inhibit the virus, an intracellular parasite, while sparing the host. Most of the early antiviral agents had a low therapeutic ratio—the ratio of the therapeutic concentration to the toxic concentration. Pharmacological problems have also been encountered. All viruses rely on the host for protein synthesis and many utilise host enzymes for transcription. Although some DNA viruses rely on host DNA polymerases, all the RNA viruses code for their own unique replicative enzymes. It is possible to identify a number of stages of the virus growth cycle which should be susceptible to selective inhibition. There are five antiviral agents in the pharmacopoeia, although only one is used to any extent. Four of these drugs were developed for the treatment of herpetic infections and the fifth for influenza.

The purine and pyrimidine nucleosides

These all act to inhibit herpesvirus DNA replication, although there is a wide spectrum of sensitivity among the different viruses. Toxic effects on the liver and bone marrow, hair loss and stomatitis are the most frequent complications of systemic therapy.

Idoxuridine (IDU)

5-iodo-2'-deoxyuridine is phosphorylated by host cell and viral thymidine kinase and acts (a) by competitively inhibiting viral DNA synthesis and (b) by incorporation into new viral DNA. As a result, both viral and cellular DNA synthesis is inhibited. However idoxuridine is too toxic for systemic use and it is rapidly denatured to an inactive form. Idoxuridine is so insoluble that even topical application to skin infections required that it was dissolved in dimethylsulphoxide (DMSO).

It was used most successfully to treat herpetic keratitis although regular installation was needed. Skin infections were treated less successfully, although many studies found that the lesions healed a day or two earlier with the application of a 5% solution in DMSO. A reduction in virus excretion was usually evident, and local treatment reduced the pain of extensive genital herpetic lesions. Similar beneficial effects were found with herpes zoster lesions, although the concentrations needed were higher (35–40%).

Cytarabine: cytisine arabinoside

This drug could be given systemically, but has now been superseded.

Vidarabine: adenine arabinoside

Until the advent of acyclovir this was the drug of choice for the systemic treatment of severe herpetic infections.

Acyclovir: Acycloguanosine

This drug is the least toxic of the nucleosides, as it has a selective inhibitory effect on herpes virus DNA polymerase. Most studies show a fairly rapid decrease in virus shedding and the development of further lesions and the time to healing is thus reduced. These effects are most marked if treatment is started as early as possible. The drug is of value in the treatment of all forms of herpes simplex infection and also varicella zoster. It has little effect on cytomegalovirus, although related compounds show an increased activity.

Acyclovir is available for intravenous infusion to treat such serious herpes simplex infections as encephalitis, disseminated infection in compromised patients, and extensive genital infection.

Oral therapy is also possible and there are preparations for the topical treatment of eye and skin infections. Despite extensive use there have been few reports of toxic effects. Acyclovir is much more soluble than idoxuridine, and is much less toxic to the cornea. The drug is most valuable in severe infections with herpes simplex and varicella zoster viruses in compromised patients, for example those undergoing renal or marrow transplants, or in AIDS. The lesions of severe primary genital infection are usually amenable to therapy. Local application of the ointment to cold sores is effective.

It must be remembered that acyclovir inhibits virus replication and has no effect on virus latent within sensory ganglia.

Resistant strains of virus have been isolated from treated patients: the most common mutation is the lack of viral thymidine kinase. In some laboratory studies resistance can be found in viral DNA polymerase. A number of other nucleosides have some effect: *trifluorothymidine*, is more soluble than IDU; *bromovinyl deoxyuridine* has improved action and less toxicity than IDU.

Other compounds

A great many compounds have been or are being investigated.

Amantadine is effective in preventing influenza virus infection if given soon after exposure. It has neurological side effects and is seldom used.

Benzimadazoles and *guanidine* inhibit some picornaviruses: derivatives have some effect on the related rhinoviruses.

Methisazone was developed to reduce the transmission of smallpox: it is highly selective for this group but is no longer relevant with the eradication of smallpox and the cessation of vaccination.

Phosphonoformate is an inhibitor of herpes simplex DNA polymase. *Ribavirin* (virazole) has a wide spectrum of activity against both NDA and RNA viruses, including respiratory syncytial virus.

Interferons

The nature and activity of interferons in inhibiting virus replication, cell division and the immune system have been described earlier. The possibility of a broad spectrum of activity combined with a relatively low level of toxicity suggested that interferons had considerable potential as therapeutic agents. Improved production techniques involving recombinant DNA technology have provided sufficient material for more extensive trials than were possible when human buffy coat cells were the main source. An antiviral effect has been shown in local eye and skin infections, respiratory infections, severe varicella infections and extensive warts and there is an inhibitory effect on hepatitis B virus replication in chronic carriers. More controversial has been the result of treatment of some tumours. Interferon is now produced by cloning techniques. Studies with the purer, more concentrated preparations now available have established that interferon does have some undesirable effects. Thus headache, malaise and fever occur within a few hours of administration of large doses. Congestion of the nasal mucosa, sore throat and increased secretion and crusting develop after some days of local treatment. Systemically, bone marrow depression, fatigue, loss of hair and confusion have been reported.

FURTHER READING

British National Formulary 1984 British Medical Association and the Pharmaceutical Society of Great Britain
Garrod L P, Lambert H L, O'Grady F 1981 Antibiotic and chemotherapy, 5th edn. Churchill Livingstone, Edinburgh
Nicholson K G 1984 Antiviral agents in clinical practice. Properties of antiviral agents. Lancet ii: 503–506
Noone P 1979 A clinician's guide to antibiotic therapy. 2nd edn, Blackwell Scientific Publications, Oxford

6

Infections of the respiratory tract

Normal flora

Although the lower respiratory tract below the cricoid cartilage is sterile in health, several species of organisms colonise the upper respiratory tract, particularly the nose and throat. These are predominantly aerobes and include species of staphylococci, streptococci including pneumococci, corynebacteria, *Haemophilus* and neisseriae.

Defences against infection (see Table 6.1)

Table 6.1 Defences against infection in the respiratory tract

Defence mechanism
Vibrissae of the nose
Broncho-constriction
The cough reflex
The mucociliary blanket

Mucosal factors
   non-specific: antitrypsin, lactoferrin, lysozyme, influenza virus inhibitors, macrophages
   specific: secretory IgA antibody

Non-specific responses   interferon, inflammatory response
Specific responses   antibody (IgA) and cell-mediated responses: sub-epithelial lymphoid follicles
Barriers to dissemination via blood   basement membrane, histiocytes

Potential pathogens (see Fig. 6.1)

Viral infections of the respiratory tract

Viral infections of the respiratory tract are very common, but

98 CLINICAL MICROBIOLOGY

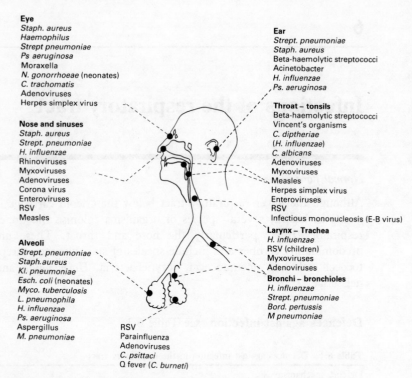

Fig. 6.1 Potential pathogens of the respiratory tract

fortunately are usually of minor significance. The most important viruses numerically are the rhinoviruses, coronaviruses, the parainfluenza viruses and respiratory syncytial virus. They may infect and re-infect at all ages to cause minor illness in adults, and have a tendency to cause descending respiratory tract infection in young children. RSV, parainfluenza and influenza viruses are the causes of severe illnesses.

Clinically it is possible to define various syndromes, for example the common cold, influenza, sore throat, croup, bronchitis, bronchiolitis and pneumonia. However, it is impossible to associate each clinical presentation with one virus as all can produce a common spectrum of illness; the extent of the illness can vary from person to person.

The common cold

Clinically the infection is of the upper respiratory tract with obstruction of the nose, clear discharge, sneezing, cough and sore throat. Fever and spread down the respiratory tract can occur. Colds are the most common type of respiratory infection and rhinoviruses are the group most often identified—upwards of 50%: coronaviruses are the cause of 10%. Many other respiratory viruses can cause a similar illness, including RSV, parainfluenza virus, influenza viruses and some echo and Coxsackie viruses.

Infection occurs at all times of the year, but in temperate climates is most frequent in the winter. Most people suffer at least one cold per season: infection occurs at all ages although it is most frequent in children. The frequency of attack is explained by the large number of viruses; more than 100 different rhinoviruses are known.

Laboratory confirmation is not usually requested, but virus isolation from respiratory secretions is the only method available.

The damage to the respiratory epithelium, the exudate and congestion can predispose to bacterial superinfection of the middle ear and sinuses.

Infection of the paranasal sinuses

Inflammation of the maxillary sinuses is common in uncomplicated virus infections of the upper respiratory tract. Bacterial infection occurs in association with obstruction resulting from mucosal oedema. This can be acute or chronic and is caused by pneumococci, *Haemophilus influenzae*, *Staphylococcus aureus*, beta-haemolytic streptococci and anaerobic cocci and bacteria that colonise adjacent areas. Material for culture can be obtained by antral lavage.

Otitis media

Otitis media is the result of direct spread of pathogens, namely pneumococci, *Haemophilus influenzae* and beta-haemolytic streptococci, from the throat via the Eustachian tube. This infection can be acute or chronic and may be caused by viruses and mycoplasmas as well as bacteria. In the young the bacteria most frequently involved are pneumococci, *Haemophilus influenzae* and beta-haemolytic streptococci, whereas in older patients *Staphylococcus aureus*, pneumococci, and in chronic cases, coliform organisms, may be responsible. Pus can be obtained by myringotomy but the eardrum may rupture naturally as the result of a build-up of pus.

Again, obstruction is an important feature in pathogenesis. Measles infection can predispose to otitis media.

Sore throat

Causes

1. Prodromal stages of infectious diseases.
2. Diphtheria.
3. Vincent's infection (see Ch. 8).
4. 'Sore throat syndrome'.

There are also non-infective causes.

Prodromal stages of systemic diseases

Many viral infections are acquired via the respiratory tract without any local symptoms. At the end of the incubation period, symptoms and signs appear reflecting infection of the target organ, for example in measles and chickenpox the mucosae of the upper respiratory tract and mouth are infected and can show a characteristic appearance before the full clinical picture develops. The appearance of Koplik's spots on the buccal mucosa in measles is a well-known example. Virus is discharged from these sites and thus the patient is infectious in the later stages of the incubation period.

Diphtheria

Bacterial aetiology

Corynebacterium diphtheriae; gravis, intermedius and *mitis* are the three main biotypes.

Pathogenesis and epidemiology

Diphtheria begins as an acute inflammatory condition of the upper respiratory tract, usually the throat, but sometimes the nasal cavity. The bacilli multiply there and produce a powerful exotoxin which diffuses through the body affecting the myocardium, adrenal glands and nerve endings. At the site of infection the inflammatory exudate and necrotic mucosal cells form a membrane that remains

INFECTIONS OF THE RESPIRATORY TRACT 101

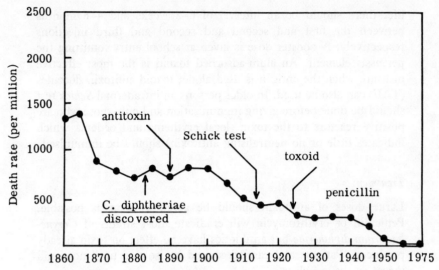

Fig. 6.2 Decline in incidence of diphtheria in the U K, 1860–1975

adherent to the throat and this may produce respiratory obstruction. Figure 6.2 shows the pattern of the disease during the last century. The carriage rate in the United Kingdom of toxigenic corynebacteria is now negligible due to the national immunisation programme.

Laboratory diagnosis

A film of a throat swab should be made to exclude infections with Vincent's organisms and the swab should be inoculated on to a Loeffler slope to allow more rapid growth of the corynebacteria, and on to a selective medium of blood and potassium tellurite. An Albert-stained smear of the growth can be made from the Loeffler slope and examined after about ten hours for the presence of metachromatic granules. Agar-gel precipitation tests (Elek) and guinea-pig inoculation tests for toxigenicity will confirm the diagnosis which is essentially clinical.

Prophylaxis

Active immunisation is the most important measure. It is best to give diphtheria and tetanus toxoids with pertussis vaccine (*triple vaccine*) in three separate doses, beginning in the third month of life; there should be an interval of 6–8 weeks and 4–6 months between the first and second and second and third injections respectively. A booster dose is given at school entry, omitting the pertussis element. An alum-adsorbed toxoid is the most effective measure when the antigen is used alone; toxoid-antitoxin floccules (TAF) can also be used. In older persons an intradermal *Schick* test should be done before giving immunisation and only those who are positive reactors to the toxin (local erythema and oedema which indicates little or no neutralising antitoxin) should be immunised.

Treatment

Large doses of antitoxin should be given as soon as possible. Penicillin or erythromycin will eradicate most strains of *Corynebacterium diphtheriae* but antibiotics have no effect on toxin already produced. Patients and carriers should be isolated until nose and throat swabs are clear.

Necrobacillosis

This rare infection caused by *Fusobacterium necrophorum* occurs in young, previously healthy adults who complain of dyspnoea and chest pain in addition to sore throat. Metronidazole is used in treatment.

The sore throat syndrome (STS)

The specific cause of the sore throat syndrome cannot be diagnosed on clinical evidence because symptoms and signs produced by both streptococci and viruses are similar, such as sore throat, pain on swallowing, tender, enlarged tonsillar lymph nodes, and pyrexia. The appearance of the throat is of little diagnostic help as non-infected throats can appear red, whereas *Streptococcus pyogenes* may be cultured from pale, normal-looking throats. An exudate over the tonsils is not necessarily an indication of streptococcal infection since many virus infections of the throat produce a similar exudate. The throat must be swabbed if an accurate diagnosis of the STS is to be made.

In viral infections, sore throat may be accompanied by the features of the common cold or of descending infection—tracheitis and bronchitis, as with influenza virus infection. Pharyngitis caused by an adenovirus may be accompanied by conjunctivitis. In closed communities, camps and long-stay homes adenovirus infection can spread widely and rapidly.

Bacterial aetiology

The STS is essentially an acute pharyngitis and/or an acute tonsillitis and is caused by bacteria, viruses and mycoplasmas. The bacteria are almost exclusively the Lancefield group A betahaemolytic streptococci, *Streptococcus pyogenes*. Groups B, C and G may also be involved, but other bacteria such as staphylococci and *Haemophilus* organisms are only very rarely implicated. For all practical purposes a *bacterial* sore throat means a *streptococcal* sore throat, except in the compromised host in whom a variety of bacteria may produce infection. It is important to remember that only 30–40% of cases of the STS are caused by streptococci; the remaining cases are most likely caused by viruses, such as adenovirus, rhinovirus, influenza, enterovirus and herpes viruses, or by *Mycoplasma pneumoniae*.

Epidemiology

Sore throat is very common especially in the winter months but it is uncommon in the under-2-year olds and in the elderly. Its peak incidence is in the young school child who succumbs to the cross-infection hazards of the classroom. It is much less common in the older school child, adolescent and young adult. Rarely a streptococcal sore throat can be contracted from eating contaminated food or drinking contaminated milk.

Laboratory diagnosis

Swabs should be sent off quickly to the laboratory because the longer the swab is in transit the greater are the chances of the organisms dying off (see Ch. 4). If delay in collection or in transit is anticipated swabs should be stored in the refrigerator as this helps preserve the streptococci that cause sore throat. A sample of saliva may also be sent in addition to the throat swab as many patients with streptococcal sore throat shed streptococci into the saliva. As *Neisseria gonorrhoeae* and *Chlamydia trachomatis* can also infect the throat these organisms may need to be looked for.

Gram films of throat swabs are not helpful unless Vincent's infection or infection with *Candida albicans* is suspected. Swabs should be plated on blood media and on crystal violet blood agar. Many beta-haemolytic streptococci are more easily recognised after anaerobic incubation of culture plates. A bacitracin disk can help to differentiate group A from other beta-haemolytic streptococci (group A is most sensitive) but streptococcal grouping tests should be performed in doubtful cases.

If virological investigation is necessary nasopharyngeal aspirate should be collected or a throat swab submitted in virus transport medium. Direct immunofluorescence gives a rapid identification of parainfluenza and influenza virus antigens; isolation takes some days. Paired sera are needed for antibody tests.

Treatment

Beta-haemolytic streptococci are universally sensitive to penicillin. Erythromycin is also a very useful drug in cases of penicillin hypersensitivity.

Acute epiglottitis: Acute laryngo-tracheo-bronchitis (croup)

Aetiology

Haemophilus influenzae, type b, is the important causative bacterium. The important viral pathogens are the parainfluenza viruses 1, 2 and 3, but other viruses, for example influenza A, B, rhinoviruses, respiratory syncytial virus and measles virus can be the cause.

Pathogenesis and epidemiology

Both conditions affect infants and young children and are frequently fulminating. Acute epiglottitis in young children due to *Haemophilus influenzae* may cause acute respiratory obstruction.

Laboratory diagnosis

In both infections there is a septicaemic phase and blood culture is frequently positive. Pharyngeal swabs may also be helpful.

Treatment

Because of the severity of both infections antibiotic therapy has to be on a 'best guess' basis, and ampicillin or chloramphenicol may be used until the laboratory confirms the diagnosis.

Bronchitis

Bacterial aetiology

Two bacteria predominate in acute exacerbations of chronic bronchitis. *Haemophilus influenzae* and *Streptococcus pneumoniae*.

Viral aetiology

Influenza, rhinoviruses and the adenoviruses are often implicated. Viral infection may precipitate an acute exacerbation of chronic bronchitis.

Pathogenesis and epidemiology

Chronic bronchitis is a relapsing condition, common in the United Kingdom. Various factors are involved in addition to infection, namely smoking, previous lung damage, atmospheric pollution and cold damp weather.

Laboratory diagnosis

Sputum samples are examined although the diagnosis is easily made clinically. One of the main reasons for culture is to check on antibiograms of the probable pathogens, and to obtain reliable results it is essential that a proper sample of sputum is taken. Sputum is homogenised prior to culture and examination of several samples of sputum is recommended.

Treatment

Each acute episode can be treated as it occurs or the patient can be put on a long-term course of antibiotics over the winter months. In the latter event it is mandatory to check periodically the antibiotic sensitivities of any *Haemophilus* organisms isolated. Broad-spectrum antibiotics such as tetracyclines and ampicillin are useful in this condition but treatment must depend on sensitivity tests.

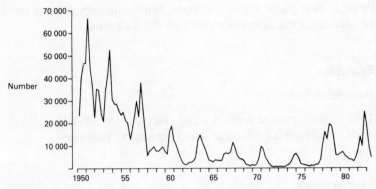

Fig. 6.3 Whooping cough: quarterly notifications, England and Wales, 1950–1982 (to the PHLS Communicable Disease Surveillance Centre, London)

Whooping cough

Bacterial aetiology

Bordetella pertussis.

Pathogenesis and epidemiology

Infection is spread by direct droplet spray and in unprotected children the secondary attack rate in a family may be around 90%. It is a serious disease in young children, especially females, and approximately 10% of children under 2 years of age who contract the disease have to be admitted to hospital. The fatality rate in children under 1 year of age is 50 times higher than in 1–4 year olds and 100 times greater than in 5–9 year olds. Immunity by infection or by immunisation is long lasting, and in so-called second attacks of whooping cough the possibility of an infection by *Bordetella parapertussis*, or by viruses should be considered. Figures 6.3 and 6.4 show whooping cough and vaccination rates in England and Wales over the last few years. Whooping cough occurs in epidemic proportions every few years.

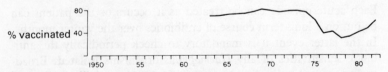

Fig. 6.4 Whooping cough: percentage of children vaccinated by end of second calendar year after birth, England and Wales, 1963–1982

Laboratory diagnosis

A pernasal swab or cough plate containing Bordet-Gengou or charcoal medium should be examined by the usual methods, and a swab may also be examined by immunofluorescence. With a good technique the organisms can be isolated from almsot every case in the early stage of the disease if the patient is not on antibiotics. As time progresses isolation becomes more difficult but by the end of the second week serological tests such as agglutination and complement fixation tests can be performed.

Prophylaxis

Antibiotic prophylaxis with, for example, erythromycin has been suggested for children at risk from the infection who cannot be immunised.

Treatment

Bordetella pertussis is sensitive to many antibiotics *in vitro* but chemotherapy is generally unsatisfactory unless given very early in the infection, probably because the organisms are usually embedded in the deeper layers of the respiratory epithelium covered by a blanket of mucus and pus, where they can withstand any antibiotic siege.

Influenza

The history of fever, aches and headaches, allied to symptoms of respiratory infection are well-recognised and have been known for centuries. Characteristically infection occurs in winter epidemics, with a significant effect on the mortality rate in the young and the elderly. Rarely, an acute viral pneumonia may develop very rapidly; more often the viral infection leads to bacterial superinfection by *Streptococcus pneumoniae*, *Haemophilus influenzae* and *Staphylococcus aureus*.

Three influenza viruses are recognised, A, B, C, although infection with the C type is seldom identified.

Pathogenesis and epidemiology

The virus is confined to the respiratory tract epithelium and causes symptoms within 1 or 2 days. The worst effects are over within a few days, although this period may be extended by bacterial invasion of the sinuses and the lower respiratory tract. Immunity to re-infection is satisfactory for some time; the response is directed against the envelope glycoproteins, the haemagglutinin (H) and neuraminidase (N) antigens. However, by mutation, minor antigenic changes occur in the H units, allowing the virus to escape the neutralising effect of antibody to earlier strains. This antigenic drift leads to regular epidemics of both A and B virus, although B viruses are more stable.

Influenza A virus also causes pandemics at irregular intervals. These major outbreaks are caused by the appearance of viruses with completely new surface antigens—the phenomenon of antigenic shift. The H and N antigens are designated by number. Table 6.2 lists the record of pandemic behaviour and the associated antigenic changes since the virus was isolated in 1933. The most severe pandemic 'Spanish flu', occurred after the first World War: the virus was related to the subsequent $H_1 N_1$ isolates.

Table 6.2 The changes in antigenic composition of influenza A

Period	Antigenic Composition
Up to 1957	H_1N_1
1957–1968	H_2N_2 Asian 'flu
1968–	H_3N_2 Hong Kong 'flu
1977–	H_1N_1

Until 1977 the appearance of a new virus led to the disappearance of the previous virus. After this shift, antigenic drift developed as expected. However this process was upset by the reappearance in 1977–78 of a virus not isolated since the 1950's. This was the first demonstration of the reappearance of H and N viral antigens. The presence of a pool of antigens had been inferred from serological studies of older people. Unlike influenza B there are influenza A viruses of birds, horses and pigs and it is believed that the H and N antigens can be exchanged into human viruses from these sources, especially birds. Acquisition of new antigens is facilitated by the fragmented genome; this property can also be used to produce new strains for vaccine production.

Control

The most severe effects of influenza are seen in the elderly and those with chronic respiratory and cardiac diseases. In these groups regular immunisation is advisable: other people may be offered vaccine for convenience and economic reasons. Killed vaccines are available (see Ch. 3). Amantadine has a selective inhibitory effect on influenza virus but has not found wide acceptance.

Laboratory diagnosis

Specimens of respiratory secretions, nasopharyngeal aspirate, throat washings or throat swabs should be submitted for direct examination for virus by immunofluorescence and for virus isolation. Serological tests require paired serum samples to demonstrate a significant rise in titre between the acute sample and another collected up to 2 weeks later.

Viral infections of the lower respiratory tract

These may develop from initial upper respiratory infection and the progression can be rapid. The most severe infections occur in young children as croup, bronchiolitis and bronchopneumonia, caused by RS virus, influenza and the parainfluenza viruses. Influenza is the most important virus in the elderly.

Both croup and bronchiolitis are serious diseases in the very young and admission to an intensive care unit may be necessary. There are annual outbreaks of both RS virus and parainfluenza virus infection. These viruses do not show antigenic variation, but infect and spread widely in the unprotected respiratory tract of young children despite the presence of maternal IgG antibodies in the blood. Vaccines are not available: indeed early attempts to protect against RS virus infection with vaccines led to enhanced disease on exposure to the natural infection.

Laboratory diagnosis

In young patients this is best attempted by immunofluorescent examination of respiratory tract cells; these are obtained from nasopharyngeal aspirate collected in a mucus extractor.

Pneumonia

Bacterial aetiology

Lobar pneumonia is invariably caused by *Streptococcus pneumoniae* whereas *bronchopneumonia* can be caused by many pathogens.

Pathogenesis and epidemiology

By definition pneumonia is inflammation and consolidation of the lung substance and can be caused by physical and chemical agents as well as a host of microorganisms. The two main infective types are lobar and bronchopneumonia, although other types such as hypostatic and aspiration are described; these names refer rather to predisposing factors than to separate pathological entities.

Cross-infection is the main means of production of lobar pneumonia, although infection may be caused by the pneumococci in the patient's own upper respiratory tract flora. There are over 80 pneumococcal serotypes but only a few are implicated in the majority of attacks. The pathogenesis involves invasion of the alveoli with subsequent deprivation of alveolar cells of adequate nutrition, resistance of the capsulate organisms to phagocytosis and the production of one or more exotoxins, such as pneumolysin.

In *bronchopneumonia* it is frequently difficult to determine the causative organism because of the multiplicity of bacteria but infection is usually endogenous and most frequently caused by *Streptococcus pneumoniae*, *Staphylococcus aureus* and *Haemophilus influenzae*. Coliform organisms are also commonly involved. The course of the disease is unpredictable and the response to treatment is frequently capricious, unlike lobar pneumonia. In the latter young adults are often affected whereas in bronchopneumonia it is the young and old who succumb.

In children bronchopneumonia may follow a viral infection of the upper respiratory tract such as measles, and in the old and debilitated staphylococcal bronchopneumonia can be a particularly serious sequel to influenza. In the latter multiple abscesses may form in the lung and the infection which is rapid and severe often has a fatal outcome.

Patients with *cystic fibrosis* are very susceptible to respiratory tract infection because of viscid bronchial secretions which reduce the efficacy of the clearing mechanisms. *Staphylococcus aureus* and *Haemophilus influenzae* commonly cause infection as well as opportunist pathogens such as *Pseudomonas aeruginosa*. Klebsiella pneumonia has a 50% mortality and occurs frequently in hospital patients with pre-existing pulmonary disease and in those on mechanical ventilation in intensive care units.

Laboratory diagnosis

A properly taken sample of sputum must be sent to the laboratory.

Blood culture may also be useful in diagnosis of lobar pneumonia. A complement fixation test is available to measure antibodies to *Mycoplasma pneumoniae* and the other organisms associated with an atypical pneumonia.

Prophylaxis

This has been attempted by the use of vaccines containing prevalent pneumococcal types, and a high degree of protection can be obtained by these. It is not a practicable proposition on a large scale but is useful for people at special risk.

Treatment

Antibiotic treatment will depend on sensitivity tests but penicillins are commonly used with a high measure of success.

Atypical pneumonia

Mycoplasma pneumoniae causes a so-called 'atypical pneumonia' in which signs of upper respiratory tract infection, fever and malaise progress to a pneumonia, with a dry cough and scanty sputum. The condition may only be suspected when initial microbial therapy fails to cure a suspected bacterial pneumonia. The organisms are sensitive to erythromycin and the tetracyclines. Isolation of the organisms is difficult and the diagnosis is made by serological tests.

Q fever

Infection with *Coxiella burneti* acquired by inhalation, presents as severe headache, malaise and aches with respiratory features developing as in mycoplasma infection. It is shed from infected animals especially sheep, cattle and goats. The organism is stable and can survive in the dried state. Some human cases may result from infected milk as the organism can survive normal heat treatment. Most infections occur in agricultural workers, vets and slaughterhouse workers. Infection is world wide and is maintained in rodents and small mammals; ticks transmit infection and may infect farm animals.

Isolation of the organism is difficult and dangerous and seldom attempted, hence diagnosis is made by serological tests. In acute infections an antibody response develops to the phase II antigen.

The organism is sensitive to tetracycline and erythromycin.
A vaccine is available and may be used for at risk groups.

Chlamydial pneumonias

These rare infections with *Chlamydia psittaci* usually occur in people with a history of contact with infected birds such as parrots. Again systemic upset is a prominent feature. Young children are also susceptible to infection with members of subgroup B of the chlamydiae. These are a major cause of genital tract infection, and can infect the newborn's eye but infection by inhalation is now recognised. The chlamydiae are sensitive to tetracyclines and some to erythromycin. Laboratory confirmation on psittacosis is by serology, whereas subgroup B infection can be diagnosed by isolation if care is taken to transport the specimen rapidly in special transport medium, or store it at $-60°C$ until inoculation. Rapid direct immunofluorescent tests are also available; ELISA may also be used.

Legionnaires' disease

Bacterial aetiology

A Gram-negative, non-acid-fast bacillus.

Pathogenesis and epidemiology

An explosive outbreak of severe respiratory illness occurred in delegates attending an American Legion Convention in Philadelphia in 1976. 29 of 182 cases were fatal. A retrospectively diagnosed outbreak also occurred in Benidorm in 1973. Sporadic cases are common. Legionellae, of which there are at least 10 serogroups, are ubiquitous. They are present in domestic and hotel water systems, in lakes, water tanks, rain forests and estuaries. The disease is contracted by inhaling aerosols of infected water from air conditioning systems which capture, concentrate and disperse the particles, from shower heads and water filters; rubber washers in shower heads and taps support growth of legionellae. The attack rate however is low.

Laboratory diagnosis

Tracheal and bronchial aspirates and needle biopsy of lung tissue are better than sputum for diagnosis. These fastidious organisms grow on enriched media producing 'cut-glass' colonies after 2–3 days. Serology however is the mainstay of diagnosis and indirect fluorescence tests are used. A titre of 256 is diagnostic.

Treatment

Erythromycin, tetracyclines and rifampicin can be used.

Respiratory tuberculosis

Bacterial aetiology

Mycobacterium tuberculosis is the causative agent; *Mycobacterium bovis* no longer causes disease in the United Kingdom.

Pathogenesis and epidemiology

The pathogenicity of tubercle bacilli depends on their ability to resist enzyme destruction when in the cytoplasm of macrophages. Intracellular growth occurs, the macrophages die and the bacilli either proceed to multiply extracellularly or, after ingestion by other macrophages, again intracellularly. Patients who recover naturally possess cell-mediated immunity and in addition usually have macrophages with an enhanced ability to kill the organisms.

The most common form of primary (childhood) infection is the production of the primary complex, comprising (1) a focus (Ghon) of infection in the subpleural area, probably at the site of implantation of the infecting bacilli; and (2) involvement of the regional mediastinal glands draining the primary focus.

In the majority of cases infection clears up spontaneously causing no symptoms but in some cases a more generalised infection may be produced either by direct spread causing tuberculous bronchopneumonia, or by the haematogenous route causing miliary tuberculosis. The latter can affect many areas including the meninges, liver, kidneys, bones and joints.

Primary infection may occur via the intestinal tract if infected milk is drunk but this now has been eradicated in the United Kingdom.

The post primary (secondary or adult) is the most common type of clinical tuberculosis. One or more lung lesions lead to caseation and cavitation and, if this involves the bronchial tree, open pulmonary tuberculosis. Infection may be exogenous but may be caused by reactivation of a healed primary lesion. Such cases must be treated speedily and vigorously with appropriate antibiotics.

The mortality from tuberculosis has fallen dramatically since the middle of the last century. Notification rates are also considerably lowered, particularly in young children and young adults, formerly two of the most susceptible groups. At the present time tuberculosis is increasingly seen as an infection of the elderly.

Laboratory diagnosis

Tubercle bacilli are not commensal organisms and to isolate them from a patient is pathognomonic of the presence of tuberculous disease. Also because commensal or saprophytic mycobacteria are rarely found in the respiratory tract the finding of even one acid-fast bacillus on a sputum smear stained by the Ziehl-Neelsen method is substantial evidence of pulmonary tuberculosis. Such a provisional diagnosis is of immense importance as it allows chemotherapy to be started instantly without having to wait several weeks for the result of culture. A good sputum sample is best, failing which laryngeal swabs or gastric washings may be examined. Fluorescence microscopy is also useful particularly when large batches of specimens have to be examined. Negative microscopical findings do not exclude the presence of tuberculosis as approximately 10^5 bacilli must be present in 1 ml sputum before positive microscopical findings can be reasonably expected.

Sputum should be cultured on egg-containing media in screw-cap containers to preserve moisture over the long incubation period of 6 weeks. Media in use are: (1) Löwenstein-Jensen (L-J); (2) modified L-J medium (IUT) which does not contain starch and is good for human type growth, although not for bovine; (3) Stonebrink's sodium pyruvate medium, which is good for bovine strains, antibiotic-resistant human strains and atypical mycobacteria.

Prophylaxis

This includes social measures, such as better nutrition. The use of selective radiography is helpful and segregation and chemotherapy of known cases is important. The use of the tuberculin test and administration of live attenuated vaccine (*BCG*) when required have also played a major role in the decrease in incidence of pulmonary tuberculosis.

The tuberculin test

The prevalence of tuberculosis can be measured by the use of the tuberculin test which when positive indicates that the person has at some time been infected by *Mycobacterium tuberculosis*, without necessarily having symptoms and signs of the disease, or has been vaccinated with BCG. Tuberculin testing indicates a delayed hypersensitivity reaction which concomitantly indicates a degree of resistance to tuberculosis. *Old tuberculin (OT)*, the specific tuberculoprotein, or a *purified protein derivative (PPD)*, is introduced into the skin of the forearm by intradermal injection (Mantoux test), or by multiple puncture (Heaf and Tine tests).

Mantoux test. 0.1 ml of 1, 10 or 100 units of OT or PPD are injected intradermally. A positive test appears as an area of induration at least 5 mm in diameter, surrounded by an area of erythema.

Heaf test. The result is read at 3–7 days and the result is regarded as positive if four or more of the six puncture sites show indurated papules at least 1 mm in diameter. Grades 1–4 indicate degrees of reaction.

Tine test. A unit of four tines (prongs) is used. A positive result consists of induration of at least 2 mm in diameter at one or more puncture sites.

Whereas positive results indicate that the person has had previous contact with infection or has had BCG vaccine and has therefore a degree of protection to tuberculosis, it can also mean that the person is currently suffering from the disease. A negative result usually excludes the disease although false negative results can be obtained in acute forms of the disease such as tuberculous meningitis or in the early stages of tuberculosis. However, strongly positive results or a positive reaction to dilute antigen suggest that the person is currently suffering from the disease.

Over the years the number of reactors has fallen markedly and it has been suggested that when the number falls to less than 1% tuberculosis will have been controlled.

BCG (Bacille Calmette-Guérin) is named after the two Frenchmen who described this vaccine in 1921. It is a bovine strain of *Mycobacterium tuberculosis* attenuated by repeated growth for many years on a bile-potato medium. The vaccine is given intradermally and should be offered to tuberculin-negative individuals who are at special risk and also to tuberculin-negative schoolchildren between 10–13 years of age. The tuberculin test should always first be performed because of the risk of severe local reactions if BCG is given to a tuberculin-positive person. Complications of vaccinations are rare.

Treatment

Various antibiotics are used, usually in combinations of three initially until sensitivities are known, and then in combinations of two. Isoniazid, para-amino salicylic acid (PAS), ethambutol, cycloserine, rifampicin, ethionamide and pyrazinamide are used. Because treatment has to be continued for many months it is mandatory that repeated sputum cultures are taken to ensure that they remain negative. There is a tendency for drug resistance to occur especially if they are used alone and this possibility must also be checked.

Lung abscess

Bacterial aetiology

Staphylococcus aureus, Streptococcus pyogenes, Streptococcus pneumoniae, various Gram-negative bacilli and anaerobic organisms such as *Bacteroides* spp, *Fusobacterium* spp and anaerobic cocci may be involved. There are usually mixed organisms present.

Pathogenesis and epidemiology

A lung abscess is usually secondary to some pre-existing condition such as pneumonia, the presence of foreign bodies, trauma and tumours. The most common group are those associated with bronchopneumonia, although they are rarer nowadays because of more effective therapy of the primary cause. Infected material is inhaled and obstructs the smaller bronchi and bronchioles producing atelectasis. Suppuration, necrosis and abscess formation occurs. Empyema or abscesses elsewhere in the body may follow.

Laboratory diagnosis

Sputum should be examined and pus should be processed if available but it is also most important to diagnose the primary cause.

Treatment

Surgery is often avoidable if antibiotic treatment is started early enough. 'Best guess' therapy can include the combination of a penicillin and metronidazole.

Aspergillosis

There are many species of the fungus *Aspergillus* and although most are of low pathogenicity they can cause pulmonary infection in patients with immunological impairment or lung tissue damage when their spores are inhaled.

When *Aspergillus* grows in a cavity (usually post-tuberculous) in the lung it produces a ball of fungus known as an aspergilloma.

Serum precipitin tests are helpful in diagnosis and the fungus can also be grown from sputum.

Empyema

This is a pleural abscess characterised by pus in the pleural space. An empyema is never primary and may be caused by: (1) extension of lobar pneumonia, bronchopneumonia, tuberculosis or lung abscess; (2) bacterial spread from the chest wall following thoracic surgery or trauma; (3) spread of infection from below the diaphragm, as in hepatic or subphrenic abscess. Pneumococci and *Haemophilus* organisms may be involved in spread from the lung, staphylococci and Gram-negative organisms from the chest wall, and coliform organisms and anaerobic cocci from under the diaphragm.

Pus must be aspirated and examined.

FURTHER READING

Christie A B 1980 Infectious diseases: epidemiology and clinical practice. Churchill Livingstone, Edinburgh

Couch R B, Kasel J A 1983 Immunity to influenza in man. Annual Review of Microbiology 37: 519–549

Garrod L P, Lambert H P, O'Grady F 1981 Antibiotic and chemotherapy, 5th edn Churchill Livingstone, Edinburgh

Horne N W 1971 Epidemiology and control of tuberculosis. British Journal of Hospital Medicine 5: 732–747

MacFarlane J T 1983 Legionnaires' disease: update. British Medical Journal 287 443–444

Ross P W 1973 Problems in the diagnosis and treatment of the sore throat syndrome. Pediatrics Digest 15: 37–44

Sudlow M S 1982 Pneumonia. Medicine International 21: 6–11

Webster R G, Laver W G, Air G M, Schild G C 1982 Molecular mechanisms of variation in influenza viruses. Nature 296: 115–121

7

Infections of the cardiovascular, blood and lymphoreticular systems

Defences against infection

Portals of entry to the bloodstream are the skin, mucosae, the genitourinary, respiratory and intestinal tracts and each has its own particular set of defences against infection (see appropriate chapters). In the bloodstream itself the defence mechanisms include the neutrophil polymorphs and macrophages from the liver, spleen, lung and marrow as well as circulating antibodies.

The effectiveness of such defence mechanisms may be upset by several factors: (1) disturbance of the normal protective flora of the respiratory, intestinal and genital tracts by the administration of broad-spectrum antibiotics; (2) variations in the immune state of the host due to diabetes, steroid therapy, immunosuppression, immune deficiency diseases, blood dyscrasias and other malignancies; (3) the ability of some organisms such as *Salmonella typhi*, *Brucella* spp and *Mycobacterium tuberculosis* to survive and multiply within macrophages. A viraemic phase is an important step in the spread of viruses from the system of entry to distant sites.

Potential pathogens (see (Fig. 7.1)

Bacteriaemia and septicaemia (see Ch. 2)

Organisms isolated from blood

The commonest isolates of clinical significance are *Staphylococcus aureus* and Gram-negative bacilli, including *Escherichia coli*, *Proteus*, *Pseudomonas*, *Klebsiella*, *Salmonella*, *Haemophilus*, *Bacteroides* and *Brucella* spp. *Streptococci*, particularly the viridans streptococci, *Streptococcus pneumoniae* and *Streptococcus faecalis* are also common isolates; neisseriae, clostridia, anaerobic cocci and *Staphylococcus albus* are less commonly found. Cytomegalovirus, hepatitis

INFECTIONS OF THE CVS AND LR SYSTEM 119

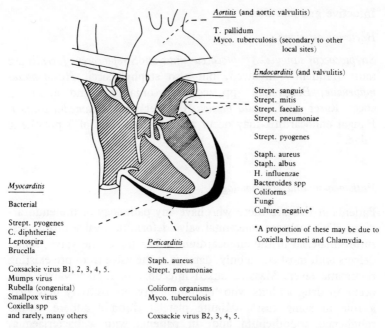

Aortitis (and aortic valvulitis)

T. pallidum
Myco. tuberculosis (secondary to other local sites)

Endocarditis (and valvulitis)

Strept. sanguis
Strept. mitis
Strept. faecalis
Strept. pneumoniae

Strept. pyogenes

Staph. aureus
Staph. albus
H. influenzae
Bacteroides spp
Coliforms
Fungi
Culture negative*

*A proportion of these may be due to Coxiella burneti and Chlamydia.

Myocarditis

Bacterial

Strept. pyogenes
C. diphtheriae
Leptospira
Brucella
Coxsackie virus B1, 2, 3, 4, 5.
Mumps virus
Rubella (congenital)
Smallpox virus
Coxiella spp
and rarely, many others

Pericarditis

Staph. aureus
Strept. pneumoniae

Coliform organisms
Myco. tuberculosis

Coxsackie virus B2, 3, 4, 5.

Fig. 7.1 Potential pathogens of the cardiovascular system

B, non-A, non-B hepatitis and human immunodeficiency viruses (HIV) can also be isolated.

Predisposing factors

Peripheral sources include injury to the skin, for example surgical wounds, intravenous tubes, injection sites, burns and graft infections.

Some viruses are injected directly into the blood via contaminated syringes and needles (hepatitis B and HIV) or by mosquitoes (togaviruses).

Central sources include therapeutic trauma such as sigmoidoscopy or even enemas, jejunal biopsy, catheterisation of the bladder, dilatation of the urethra and dental extraction. Collections of pus under pressure and infected tumours are also potential sources. In the patient receiving immunosuppressive drugs, cytotoxic drugs or corticosteroids, the immune mechanisms are less able to cope with invading organisms and organisms of lesser pathogenicity can produce septicaemia. Also, in the neonate with a congenital deformity the chances of septicaemia are high.

Infective endocarditis

Bacterial aetiology

Streptococcus sanguis, Streptococcus mitis and *Streptococcus faecalis* are most frequently involved, but also staphylococci, *Haemophilus influenzae, Bacteroides* spp, coliform organisms and anaerobic cocci. Rarer are *Coxiella burneti* (Q fever) and *Chlamydia psittaci*. Fungal endocarditis may occur after the insertion of a prosthetic valve.

Pathogenesis and epidemiology

Patients at risk are those who have any pathology of the endocardium. This includes congenital valve deformities, valve prostheses, suture material in the endocardium, atheroma of the valve, septal defects and, most commonly, damage to the valve from pre-existing rheumatic fever. Massive mixed infections of normal valves may occur in drug addicts who 'mainline'. Hypersensitivity may play a role in some cases. Minute platelet thrombi adhere to the roughened endothelium and, in patients with a bacteriaemia, bacteria may grow in the thrombi and produce fragile vegetations from which emboli can break off. The vegetations consist of fibrin and platelets and the bacteria are centrally placed in these. Fibrosis can occur in the valve and this surrounds and protects the bacteria even more extensively.

Acute and subacute forms can occur; pyogenic organisms are more common in the acute form. Over the years the major change in the pathogenesis has been a reduction in the number of cases caused by the viridans group of streptococci and an increase in those caused by the enterococci and staphylococci. Also, whereas the age group commonly involved was up to and around 30 years of age, nowadays infective endocarditis is becoming more and more a disease of the elderly and in these a previous history of rheumatic fever is still very common.

Laboratory examination

This is one of the many infections where close collaboration between ward and laboratory is vital in diagnosis and in the control of antimicrobial chemotherapy. Blood culture is the main diagnostic procedure. Occasionally blood culture results may be equivocal or give negative results during the first few days if organisms grow slowly. The course of endocarditis due to *Coxiella burneti* is much more insidious although eventually valve damage is severe. There is seldom a clear history of preceding respiratory tract infection. The mortality rate in blood culture-negative endocarditis exceeds that of cases where an organism has been identified.

Blood culture

Bacteria can be isolated from first cultures in more than 80% of patients with infective endocarditis and bacteriaemia is often more persistent than intermittent. Many media can be used for culture of blood, such as cooked meat broth, brain heart infusion and thiol broth. It is important to subculture on to solid media and incubate both aerobically and anaerobically.

Culture-negative endocarditis

Several reasons are suggested for this:
1. Chemotherapy has been started.
2. Bacteriaemia is intermittent and inadequate numbers of specimens have been taken.
3. Organisms may be fastidious or slow growers like *Bacteroides* spp; these require proper media and an adequate period of incubation. Nutritionally-variant strains, for example pyridoxine-dependent *Streptococcus mitis*, may be the causative agents and these will not grow on standard media.
4. *Coxiella burneti* or chlamydiae may be involved: establishing the diagnosis is important so that appropriate treatment may be started. In suspected cases, the diagnosis is established by serological tests. Complement fixation tests are widely used and examination of a single serum will be diagnostic if raised titres are found to both phase I and II antigens. Titres are usually greater than 256 and are often much higher. Chlamydial infection is much rarer: again serological tests are needed to establish the diagnosis.
5. Occasionally organisms that may be reported as 'contaminants', for example *Staphylococcus epidermidis* (*albus*), may be the causative organisms.

Prophylaxis of infective endocarditis

This means antibiotic cover for susceptible persons during dental extractions, urinary tract and intestinal procedures and cardiac surgery.

For dental extraction susceptible persons should be given penicillin, preferably by *injection*, about 30 minutes before extraction. Penicillin with streptomycin or gentamicin may be given if culture of saliva has revealed penicillin-resistant viridans streptococci and patients with prosthetic valves should be given such a combination when undergoing dental treatment. Oral penicillin may be used if injections are refused or difficult to arrange and oral erythromycin may be substituted in penicillin-hypersensitive patients, started one hour before extraction, and, as with oral penicillin, continued for 24 hours.

Penicillin with streptomycin or gentamicin, or ampicillin with gentamicin are very suitable combinations in those undergoing urinary or intestinal tract manipulations because of the effect against *Streptococcus faecalis*.

In those about to undergo cardiac surgery a combination of benzyl penicillin or flucloxacillin and an aminoglycoside such as gentamicin will be effective against most relevant pathogenic bacteria.

Treatment

Laboratory monitoring of therapy is crucial and measurement of the MIC and MBC of antibiotics for the isolated organisms is required, particularly to ensure that serum levels are adequate to penetrate the valves and kill the bacteria. Doses have often to be very high to ensure this. It is essential that parenteral antibiotics are used at least for the first 2 weeks of treatment. There are reports of high levels of resistance to streptomycin in some strains of *Streptococcus faecalis*. Combined therapy may be more effective for all viridans streptococci as well as for the enterococci, for example a penicillin and an aminoglycoside; ampicillin or amoxycillin may be preferable to benzyl penicillin and in most cases gentamicin is the best choice of aminoglycoside. In all cases treatment must be continued for at least 6 weeks. In *Coxiella burneti* infection surgical excision of the valve is usually required with long term treatment with tetracyclines or clindamycin.

If possible, treatment should start before the operation and continue through the operative period and beyond. Over several years, the serum antibody titres will fall gradually: treatment may then be stopped.

Rheumatic fever (RF)

Bacterial aetiology

Streptococcus pyogenes.

Pathogenesis and epidemiology

Two main theories have been proposed to explain the occurrence of this non-suppurative streptococcal complication: (1) that the initiation of the chronic lesions in the heart is due to the effect of some of the diffusible products of *Streptococcus pyogenes*, such as streptolysins O and S and proteinase; and (2) *the more accepted theory*, that immunological phenomena such as immune-complex disease, cross-reactive immunity or delayed hypersensitivity develop in certain persons who become sensitised to one or more streptococcal products. An immunological relationship exists between streptococcal antigens and human myocardial tissue and one such antigen responsible for cross-reactivity is localised in the cell wall and is associated with the M-protein. It is probable that several cross-reactive antigen-antibody systems occur between the streptococcus and heart tissue, and according to the currently-accepted theory cross-reactive antibody occurs to heart valves as well as to heart muscle, group-specific polysaccharide antigen of *Streptococcus pyogenes* having been found to cross-react with the structural glycoprotein of both human and bovine valves. Specific host factors have not been excluded in the pathogenesis of this disease. As well as the above examples of the connection between *Streptococcus pyogenes* and RF there is other evidence to associate the two, namely *clinical, bacteriological, epidemiological, serological* and *chemoprophylactic.*

Clinical

Physicians in an earlier era observed the common occurrence of RF a few weeks after an attack of scarlet fever and on many occasions patients admit to having had a sore throat several weeks before the onset of RF.

Bacteriological

Streptococcus pyogenes can often be isolated from the throat of RF patients.

Epidemiological

Altitude, humidity and other physical factors were all thought to have a role in the production of RF but it would now appear that RF can occur anywhere in the world. In general, rates increase when the streptococcal carriage rate increases and in the United Kingdom the age group which has the highest incidence of RF (7-9) is also that which has the highest incidence of streptococcal sore throat.

Serological

RF does not occur in the absence of immunological evidence of streptococcal infection. The first antibody to be measured was antistreptolysin O (ASO); titres become detectable in the second week after the onset of infection, are maximal by the sixth week and thereafter decrease. Other antibody levels can now be measured, such as antihyaluronidase (ASH), antistreptokinase (ASK), antideoxyribonuclease B (Anti DNAse B) and antinicotinamide adenine dinucleotidase (anti NADase). Most laboratories perform only the ASO and sometimes the anti DNAse B tests.

Chemoprophylaxis

There is proof that penicillin prophylaxis after a primary attack of RF will reduce substantially the risk of a second attack, by preventing further streptococcal infection of the throat.

Laboratory diagnosis

Throat swabs may or may not yield growth of *Streptococcus pyogenes* and so the most important investigations are serological. Two samples of serum must be assayed to check for a rising titre of antibody to *Streptococcus pyogenes* because a single estimation does not indicate whether a raised level of antibody pertains to a current infection or to one contracted several weeks or months previously.

Prophylaxis

The incidence of *primary attacks* is less than 1% in the United Kingdom. To prevent these is virtually impossible. Penicillin treatment of streptococcal sore throat has been said to be of value in prevention but there is a great deal of evidence to show that this plays little part. For example, the incidence of RF in the United Kingdom had greatly decreased before the advent of antibiotics and, also, only a fraction of the population with streptococcal infection is treated with antibiotics because infection may be subclinical. In addition, the majority of persons with streptococcal sore throat tend not to visit their doctor for treatment. Prevention of second or subsequent attacks of RF is quite a different matter because the person, usually a child, who has had a primary attack of RF is prone to further attacks and daily prophylaxis using oral penicillin, or erythromycin in cases of hypersensitivity to penicillin, undoubtedly has proved of great benefit.

Treatment

Benzyl or phenoxymethyl penicillin are the drugs of choice in non-hypersensitive patients in order to eliminate streptococci. Erythromycin is also useful.

Acute pericarditis

The most common causes of pericarditis are viruses, notably Coxsackie B viruses. In general myocarditis and pericarditis occur together as an acute illness. The syndrome was first recognised in the newborn as part of a generalised infection. However there is evidence to support their role in infections in older patients. Virological diagnosis is difficult in the adult as serum antibody titres can remain high (greater than 512) for years after an infection at any site. Virus isolation from faeces should also be attempted. Bacterial causes are not common but *Staphylococccus aureus*, pneumococci and neisseriae may be involved either by direct spread from intrathoracic infection or as a complication of septicaemia. *Mycobacterium tuberculosis* may very occasionally be associated with the condition. Pericarditis is also a component of the rheumatic carditis syndrome.

Blood vessels

Damage to blood vessels can occur by deposition of infected emboli or immune complexes. Some cases of polyarteritis nodosa are associated with the deposition of complexes containing hepatitis B surface antigen.

Malaria

Plasmodium vivax is the most common organism but other species such as *falciparum, ovale* and *malariae* also cause different types of malaria.

All *Plasmodium* species have a sexual multiplication stage in the female anopheline mosquito and an asexual multiplication stage in man: the incubation period varies from 6–18 days.

Laboratory diagnosis

Blood films should be stained by Leishman's or Giemsa's stains and examined for plasmodia; typical ring forms may be seen.

Treatment

Depending on the type of malaria chloroquine or primaquine can be used. Quinine may also prove useful.

Virus infections of blood and lymphoreticular system

Erythrocytes

These are seldom involved in the pathogenesis of virus infections. Colorado tick fever is caused by a togavirus which circulates in the erythrocytes of the human. Anaemia is a feature of some *Mycoplasma pneumoniae* infections, probably due to the development of a cold agglutinin for human group O cells during the disease. A cold agglutinin is also recognised in infectious mononucleosis. A special association has been shown between a human parvovirus and transient aplastic crises in children with sickle cell disease and other haemoglobinopathies. The crisis lasts for some 5–10 days, and many are preceded by a febrile illness. A similar entity has been recorded in patients with hereditary spherocytosis, pyruvate-kinase deficiency, β-thalassaemia and some acquired haemolytic anaemias. In the laboratory the virus can inhibit the growth of late erythroid progenitors. Clinically purpura resulting from loss of platelets has also been recognised in acute parvovirus infection.

Platelets

Post-infectious thrombocytopaenia occurs after rubella and varicella infection and is seen in HIV, Lassa & Marburg virus infections.

The lymphoreticular system

This is an important defence against the entry and spread of infectious agents, although in some viral infections this system is an important site of virus replication. *Lymphadenopathy* is a common finding in infection, usually related to the site of bacterial infection. In some viral infections localised adenopathy is found, for example the occipital and preauricular involvement in rubella. Cervical adenopathy is a feature of glandular fever and the persistent generalised lymphadenopathy associated with infection by HIV and is sometimes seen in cytomegalovirus infection and toxoplasmosis. Compromised patients who experience numerous infections will have adenopathy at many sites. The extreme example of this is the acquired immune-deficiency syndrome (AIDS).

Glandular fever or infectious mononucleosis

This is an infection of young adults who present with fever, aches, sore throat often with exudate, a macular rash (especially if given ampicillin), local or generalised adenopathy, and often splenomegaly. Biochemical evidence of hepatitis is present and in 5% of cases is severe enough to cause mild jaundice. The causal agent is the Epstein-Barr virus (EBV) first detected in lymphoblastoid cells cultured from the lymphomas described by Burkitt in African children. The virus has an important role in the development both of this tumour and of nasopharyngeal carcinoma in Chinese. Infection with the virus is world-wide and usually unrecognised unless acquired in older children and adults. The virus has the appearance of a typical herpes virus. It can only be grown in B lymphocytes in the laboratory although it may infect epithelial cells *in vivo* in the mouth and duct cells of the parotid glands. As with other herpes viruses the virus persists after the initial infection. Virus can be recovered from such lymphocytes by *in vitro* cultivation of the cells. The role of the virus in tumour production is not fully understood. These tumours are always monoclonal in origin in contrast to the polyclonal, oligoclonal or monoclonal B-cell lymphomas seen in young men with a familial X-linked disorder and in patients who are immunosuppressed for transplantation, or as a result of infection with HIV.

Virus excreted in the saliva is spread by close mucosal contact—hence the alternative name the 'kissing disease'. The incubation period is 4–7 weeks, and symptoms are worst for 2–3 weeks, although some patients suffer symptoms for some months.

Diagnosis

Examination of the peripheral blood shows a relative or absolute lymphocytosis, and there are large numbers of enlarged lymphocytes with distorted nuclei. B cells are infected, and Tc cells react and are present in increased numbers. Numerous irrelevant immune responses are activated, including the production of a heterophil antibody to sheep or horse erythrocytes. The detection of this IgM antibody (in the Paul-Bunnell test) is diagnostic of glandular fever, provided the correct absorptions of the sera are included in the test. The Paul-Bunnell test is most useful in older patients and is indicative of recent infection; it is usually positive when the patient presents although some patients have symptoms for up to 5 days before the antibody appears. In children the test may never be positive. A specific serological diagnosis is made by immunofluorescence tests of serum to detect IgG and IgM antibodies to viral antigens expressed on carrier cultures of lymphoblastoid cells. Antibody to the virion capsid antigen (VCA) is the most widely used, but responses to other antigens may be of diagnostic value in some patients.

Human T lymphotropic viruses (HTLVs)

There HTLVs, which are human retroviruses, are presently recognised. The first, HTLV-1 is the cause of adult T-cell leukaemia-lymphoma recognised in Japan, the Caribbean and some other parts of the world. HTLV-II was isolated from a case of hairy cell leukaemia. The third, human immunodeficiency virus (HIV) otherwise known as HTLV-III or lymphadenopathy associated virus (LAV) has been isolated from patients with AIDS or its precursor states. All of these viruses infect T lymphocytes; HTLV-

I and II cause proliferation, while HIV is cytolytic. This results in a loss of T helper cells and a marked predisposition to opportunistic infections such as *Pneumocystis carinii* and *Candida albicans*. Clinically infection may present as persistent generalised lymphadenopathy (PGL) or prolonged fever, diarrhoea and weight loss (the AIDS-related complex or ARC).

Groups known to be at increased risk are residents or visitors from several African countries, including Zaire and Uganda, male homosexuals, intravenous drug abusers and recipients of blood and especially blood products derived from large donor pools.

However, heterosexual contact is also an effective means of spread and transfer in this way is known to have occurred from bisexual men to their partners, from drug abusers to their partners and from haemophiliacs to their spouses. Children have a 50% risk of infection if their mother is a carrier.

The disease was first recognised in the United States in 1981 but is now world-wide. Retrospective studies show that the virus has been present in San Francisco homosexuals and in haemophiliacs since 1978.

In areas of central Africa between 5–7% of the population have antibodies to the virus, in urban areas, infection is seen in promiscuous heterosexuals.

Transmission of HIV virus occurs by the same routes as hepatitis B and the risk groups are very similar. Blood is a source and saliva and semen have been shown to contain the virus within lymphocytes. Virus has also been demonstrated in saliva and tears. There is now evidence that the virus can infect cells in the brain and this has been linked to the progressive dementia seen in patients with AIDS. The sequence of events after inoculation is uncertain, but from studies of accidental transmission, the acute stage of HIV infection can result in fever and rash. Antibodies to the virus develop within a few weeks, although AIDS may take up 4–5 years to develop and incubation periods of only a few months have been recorded in children. The virus persists despite the presence of antibody and it appears that all patients with antibody are infectious. To date approximately 1 in 10 infected individuals have developed AIDS, more have presented with PGL and ARC, but approximately two thirds are well. The prevalence of infection in different risk groups varies considerably in different areas. Thus the majority of promiscuous male homosexuals are now infected in San Francisco, whereas in London the proportion is 25% in similar patients with other sexually transmitted diseases. Infection of haemophiliacs is now very extensive, particularly if commercial factor VIII has been used. Infection in intravenous drug abusers is recognised in the eastern states of America, and in Europe, especially Spain, Italy and Switzerland. In the United Kingdom antibody prevalences of more than 33% have been found in similar patients in Edinburgh. In the blood donor population in the United Kingdom the carriage rate of antibody is less than 1 in 20 000.

Diagnosis and control

Detection of antibody to HIV membrane antigens is the only screening test available. To date this has been used for the confirmation of the association with the disease and for epidemiological studies in risk groups. The test may fail to detect infected individuals and tests for virus isolation, viral antigens or nucleic acid will be necessary to assess the prognosis in individual patients.

Prevention by the use of a vaccine would be ideal but the prospect of the development of this is complicated by the considerable heterogeneity already detected in HIV isolates.

Passive immunisation is not available. Various other chemotherapeutic agents are on trial: most are inhibitors of the viral reverse transcriptase. If effective, longterm treatment may slow or prevent the progression to AIDS, although elimination of the virus may not be possible.

Considerable educational and counselling services are needed to slow the spread of this infection and to support those who are infected. Because the disease runs a protracted course, considerable resources will be needed.

Infectious agents in AIDS

Protozoa	Pneumocystis carinii — pneumonia
	Toxoplasma gondii — encephalitis
	Cryptosporidium — chronic diarrhoea
Fungi, yeasts etc	Candida albicans — oral, oesophageal
	Cryptococcus neoformans — meningitis
	Aspergillus — pneumonia, disseminated
Bacteria	Listeria monocytogenes — meningitis
	Invasive Salmonella infection
	Mycobacteria, including avian strains

Viruses	Reactivated infection: herpes simplex 1 and 2, varicella zoster and cytomegalovirus. These may be present as severe localised or generalised infections.
	Papovaviruses — warts, PMLE
	Adenoviruses
Tumours	Kaposi's sarcoma
	B-cell lymphomas

FURTHER READING

Banatvala J E 1983 Coxsackie B virus infections in cardiac disease. In: Waterson A P (ed) Recent advances in clinical virology. Churchill Livingstone, Edinburgh, pp 99–115

Cawson R 1982 The nature and prevention of bacterial endocarditis. Medicine Publishing Foundation Symposium Series 3. Medicine Publishing Foundation, Oxford

Haverkorn M J 1974 Streptococcal disease and the community. Excerpta Medica, Amsterdam

Hayward G W 1973 Infective endocarditis: a changing disease. British Medical Journal 2: 706–709; 764–766

Klein G, Klein E 1984 The changing face of EBV research. Progress in Medical Virology 30: 87–106

Oakley C M 1980 Infective endocarditis. British Journal of Hospital Medicine 24: 232–243

Shanson D C, Kirk N, Humphrey R 1985 Clinical evaluation of a fluorescent antibody test for the serological diagnosis of streptococcal endocarditis. Journal of Clinical Pathology 38: 92–98

Wannamaker L W, Matsen J M 1972 Streptococci and streptococcal diseases. Academic Press, London

Watt B 1978 Streptococcal endocarditis: a penicillin alone or a penicillin with an aminoglycoside? Journal of Antimicrobial Chemotherapy 4: 107–109

Wong-Staal F, Gallo R C 1985 Human T-lymphotropic retroviruses. Nature 317: 395–403

8

Infections of the mouth

Normal flora

The mouth contains many different microorganisms in an ecosystem of great complexity, much of which is not yet fully

Table 8.1 The normal flora of the mouth

(i) *Organisms generally present that constitute a major fraction of the oral flora*
Viridans streptococci (*Streptococcus salivarius, sanguis, mutans, milleri, mitis*)
Corynebacterium spp
Neisseria spp
Haemophilus spp
Veillonella spp
Lactobacillus spp
Leptotrichia spp
Actinomyces spp
Bacteroides spp
Bacterionema spp
Eikenella spp

(ii) *Organisms generally present that constitute a minor fraction of the oral flora*
Anaerobic cocci
Streptococcus pneumoniae
Staphylococcus spp
Fusobacterium spp
Spirochaetes (*Borrelia* and *Treponema*)
Campylobacter spp
Clostridium spp
Mycoplasma spp
Candida spp
Nocardia spp
Actinomyces spp
Propionibacterium spp

(iii) *Organisms that are transient or associated with the carrier state in the mouth*
Beta-haemolytic streptococci
Staphylococcus spp
Neisseria meningitidis
Coliform organisms
Corynebacterium diphtheriae
Protozoa
Viruses

understood. It is now known that the oral cavity is not a single habitat and that there are quite separate areas and sites in which microorganisms multiply. These are: the gingival crevice; the teeth; the saliva; and the mucosal surfaces, including the tongue.

Table 8.1 lists the more common organisms found in the oral cavity. In addition to these are spirochaetes (*Borrelia*), yeasts, *Mycoplasma* (*orale* and *salivarium*) and protozoa (*Entamoeba gingivalis* and *Trichomonas tenax*). Viruses may also be isolated but they are not part of the normal oral flora.

Development of the oral flora

Birth and neonatal period

The mouth of the newborn baby is usually sterile but it rapidly acquires organisms from the mother and also from the environment. Several streptococcal and staphylococcal species may be isolated, together with coliforms, lactobacilli *Bacillus* spp, *Neisseria* spp and yeasts. The selectivity of the mouth as an environment is demonstrated even at this time because most of the organisms introduced fail to become established. *Streptococcus salivarius* is the most common isolate from the mouths of young babies together with *Staphylococus albus*, *Neisseria* spp and *Veillonella* spp.

Infancy and early childhood

The infant comes into contact with a wide range of microorganisms and some of these will become established as part of the commensal flora of the individual. The eruption of deciduous teeth provides a different surface for microbial attachment and this is characterised by the appearance of *Streptococcus sanguis* and *mutans* as regular inhabitants of the oral cavity. A few anaerobes become established but as there is no deep gingival crevice the numbers remain small. Actinomycetes, lactobacilli and *Rothia* are found regularly.

Adolescence

The greatest increase in numbers of organisms in the mouth occurs when the permanent teeth erupt. The gingival crevice is deeper than in the deciduous dentition and allows for a great increase in anaerobic organisms. *Bacteroides* spp become established in large numbers as well as *Leptotrichia* spp, *Fusobacterium* spp and spirochaetes.

Adulthood

Most studies of the adult oral flora show that considerable variation occurs among individuals in total numbers and proportions of many species of bacteria; indeed there may be variation within one individual when sampled on several occasions. The adult flora is extremely complex.

Consistent with the trends seen in adolescence there is an increase in *Bacteroides* spp and spirochaetes with advancing periodontal disease and maturity of dental plaque. Superficial plaque contains many streptococci, mostly *Streptococcus mutans*, *mitior* (*mitis*) and *sanguis*. Actinomycetes and other Gram-positive and Gram-negative filaments of uncertain taxonomic position are also regularly isolated.

Defences against infection

The oral cavity represents a very sophisticated ecosystem and there are many factors that contribute to the prevention of infection (see Table 8.2).

In addition to these specific factors there are also general factors such as dental hygiene, the presence or absence of fluoride in

Table 8.2 Defences against infection in the mouth

Anatomical factors
Morphology of the teeth and their spatial relationship
Intact keratinised oral mucosa
Tongue movements
Salivary flow and swallowing reflex

Salivary factors
Lysozyme from salivary glands, crevicular exudate and polymorphonuclear
  leucocytes
Thiocyanate-dependent factors
Lactoferrin
pH
Glucolytic activity of epithelial cells
Microbial antagonism
Normal flora
Salivary glycoprotein
Leucocytes from the gingival crevice
Immunoglobulins (IgA and secretory IgA, IgG and IgM from the salivary glands;
  IgG, IgA and IgM from the crevicular exudate)

drinking water, the nutritional status and general health of the host.

Potential pathogens (see Table 8.3)

DENTAL CARIES

Bacterial aetiology

Bacteria, a susceptible host and a cariogenic diet are the three essentials for the production of caries. Viridans streptococci, chiefly *Streptococcus mutans* and to a lesser extent *Streptococcus sanguis*, are closely associated with the production of caries and are early colonisers of teeth. Other acidogenic bacteria including *Actinomyces* and *Lactobacillus* spp may contribute to the progression of caries, but are far less important than the viridans streptococci.

Pathogenesis and epidemiology

Rats and hamsters with a normal oral flora develop caries when fed on a diet rich in sucrose, whereas germ-free animals fed on the same diet remain free from caries. These animals however readily develop lesions when infected orally with a pure culture of *Streptococcus mutans*.

Plaque

This is a soft, often whitish, adherent deposit that forms on a tooth surface. Although plaque forms particularly after the abandonment of dental hygiene it is a constant factor in all mouths. It comprises a mucopolysaccharide matrix in which extracellular material, squamous cells and bacteria are embedded. One of the outstanding features of plaque is its dynamic nature. Sites on the same tooth, only 1 mm apart for example, can differ markedly in microbial composition. Also the quantity and nature of plaque on the same tooth can vary. Calcification of plaque produces *calculus*. *Supragingival plaque* is associated with *dental caries*, *subgingival* with *periodontal disease*.

Formation of dental plaque

The initial thin film, or pellicle, that coats the surface of the tooth consists of glycoproteins from the saliva which are rapidly adsorbed on to the teeth.

Early colonisation

The pellicle first builds up in defects or pits on the enamel surface and then spreads over the smooth surface of the tooth. Bacteria are associated almost immediately with the pellicle, as early as within 1 hour of its formation; species most commonly found are streptococci, neisseriae, lactobacilli and actinomyces. After a few days anaerobes, notably veillonellae appear. By the end of a week the numbers of streptococci and neisseriae decline in relative terms, accompanied by a proliferation of anaerobes such as fusobacteria and bacteroides.

7–14 days

Plaque at this stage is becoming mature and there is great variation in its composition because of microbial interactions. Virtually no streptococci or neisseriae remain and filamentous forms begin to invade the deeper layers. Prominent groups of organisms are bacteroides, actinomycetes, bacterionema, fusobacteria, leptotrichia, spirochaetes and protozoa.

14–21 days

Anaerobes and spirochaetes predominate at this stage, most other organisms having died off due to insufficient nutrition and the production of toxic substances.

Production of dental caries

There are two main theories of caries production: (1) acidogenic and (2) proteolytic. The *acidogenic theory*, the more probable, postulates that the initial lesion is a breach of the enamel caused by products of bacterial metbolism, notably of *Streptococcus mutans* and *Streptococcus sanguis*. These streptococci produce large amounts of extracellular polysaccharide, particularly dextran from sucrose, and this enables bacteria to adhere to the teeth. Other bacteria can thereafter contribute to the progression of caries. These include other viridans streptococci, which can store intracellular polysaccharide when there are sufficient carbohydrate supplies, lactobacilli, *Propionibacterium, Actinomyces* and *Arachnia*. When these become scarce the intracellular stores can be utilised ensuring continued production of acids. The *proteolytic theory* states that proteolysis is the initial event, followed by decalcification.

A third theory put forward is the *proteolysis-chelation* theory which postulates primary proteolytic digestion of the organic matrix with subsequent decalcification by chelation of calcium with products of proteolytic digestion, such as aminoacids.

Prophylaxis

Prophylaxis of caries includes general measues such as brushing the teeth, proper diet and fluoridation of water supplies. In recent years there have been attempts to produce an effective vaccine, based on the belief that carious lesions are most likely to be initiated by *Streptococcus mutans*. Serotype C of this organism has been used in studies in animals and this strain produces antibodies that can reach the sites where caries is likely to develop and exert a protective effect. Problems exist in the use of such a vaccine, because if *Streptococcus mutans* is not the only organism associated with caries the vaccine would not protect against the action of other organisms. Also, the vaccine is not sufficiently pure for parenteral use in humans; cross-reactivity between heart muscle and antibody has been reported.

Treatment

Antibiotics have no place in the treatment of caries and once the lesion is established reparative dental treatment is required.

PERIODONTAL DISEASE

Bacterial aetiology

In recent years identification of organisms involved in periodontal disease has become more firmly based and the association of a group of organisms, rather than one or two particular species, is now recognised. These include *Actinomyces, Fusobacterium, Bacteroides, Leptotrichia*, spirochaetes, vibrios, viridans streptococci, anaerobic cocci and *Campylobacter*.

Table 8.3 Infections of the Mouth

Infection	Associated organism(s)
(1) INFECTIONS RELATED TO BONE AND/OR TEETH	
Periodontal (dental) abscess	Viridans streptococci and anerobes
Osteomyelitis	Most commonly *Staphylococcus aureus*
(2) BACTERIAL INFECTIONS OF THE SOFT TISSUES	
Acute	
Streptococcal stomatitis and streptococcal sore throat	Beta-haemolytic streptococci (mostly Lancefield's Group A)
Diphtheria	*Corynebacterium diphtheriae*
Acute necrotising ulcerative gingivitis and Vincent's angina	Fuso-spirochaetal organisms, *Bacteroides melaninogenicus* and *gingivalis*
Gonorrhoea	*Neisseria gonorrhoeae*
Primary syphilis	*Treponema pallidum*
Acute sialadenitis	*Staphylococcus aureus*, beta-haemolytic streptococci
Chronic	
Actinomycosis	*Actinomyces israelii*
Cancrum oris (noma)	Fuso-spirochaetal organisms and *Bacteroides* spp.
Secondary and tertiary syphilis, bejel	*Treponema pallidum*
Tuberculosis	*Mycobacterium tuberculosis*
Leprosy	*Mycobacterium leprae*
Chronic sialadenitis	Oral streptococci and *Haemophilus influenzae*
(3) VIRAL INFECTIONS OF THE SOFT TISSUES	
Local	
Herpetic stomatitis and cold sores	Herpes simplex virus
'Shingles' or herpes zoster	Varicella zoster virus
Sore throat	Adenovirus and other respiratory viruses
Herpangina	Coxsackie A virus
As manifestations of systemic infection	
Measles	Measles virus
Mumps	Mumps virus
Chickenpox	Varicella zoster virus
Molluscum contagiosum	Molluscum virus
Hand, foot and mouth disease	Coxsackie A virus
Glandular fever	
Burkitt's lymphoma	Epstein-Barr virus
Rabies	Rhabdovirus
(4) FUNGAL INFECTIONS OF THE SOFT TISSUES	
Candidiasis	*Candida albicans*
Histoplasmosis	*Histoplasma capsulatum*
South American blastomycosis	*Paracoccidioides brasiliensis*
Coccidiomycosis	*Coccidioides inimitis*
Sporotrichosis	*Sporotrichum schenkii*

Pathogenesis and epidemiology

The periodontium comprises the gingivae, alveolar bone and periodontal membrane. Periodontal disease includes disease of any of the components and is a common human problem, particularly after adolescence. *Marginal gingivitis* is one of the commonest forms of periodontal disease and this may be acute or chronic.

The typical inflammatory response of periodontal tissues does not occur in the absence of bacteria and clinical and epidemiological observations, as well as animal studies, have provided firm evidence of the aetiological role of *subgingival plaque* in this disease.

Periodontal disease is produced by a combination of two main processes: (1) the effect on the tissues of bacterial products such as collagenase, hyaluronidase and neuraminidase, inducing an inflammatory reaction; and (2) the consequent production of immune responses, both humoral and cell-mediated, in the host.

Prophylaxis

Oral hygiene is of paramount importance in the prevention of this desease.

Treatment

Only in the most acute cases does antibiotic therapy have a place in the treatment of periodontal disease.

Pulp and periapical infections

Bacterial aetiology

Almost invariably infections of pulp and acute periapical abscesses are endogenous and many commensal bacteria may be involved; these are most commonly viridans streptococci and anaerobic cocci, staphylococci, *Neisseria*, *Haemophilus*, *Actinomyces*, *Fusobacterium*, *Bacteroides* and *Veillonella*, although it is often difficult to prove which particular species, or combinations of organisms, have actually produced the infection.

Pathogenesis and epidemiology

Infection of the pulp and periapical areas can occur by three pathways: (1) through the tubules of dentine, associated with caries; (2) by invasion along the periodontal membrane, associated with periodontal disease; and (3) through the bloodstream. Acute or chronic pulpitis can occur; in the acute form the inflammatory exudate in the pulp chamber can produce a rise in pressure causing spread of infection to the periapical tissues, either as cellulitis or as acute periapical abscess; a chronic lesion may result.

Laboratory diagnosis

Pus samples can be taken if practicable, and samples can also be taken from the root canal during operative procedures, using sterile paper points.

Treatment

Antibiotic therapy may be used to good effect in acute cases if drainage of the periapical area cannot be achieved. Once drainage has been established proper preparation of the root canal and adequate root filling measures are of paramount importance.

COMMON ORAL INFECTIONS

Oral candidiasis

Infections with *Candida* spp, including *Candida albicans*, *Candida stellatoidea* and *Candida tropicalis*, occur in both neonates and the elderly. In neonates the infection ('thrush') is acute and is contracted from the mother's genital tract. It is characterised by flaky, loosely-adherent membranes that cover the tongue, cheek and gums. Cross infection is common in neonatal nurseries.

In the elderly patient there is frequently some predisposing cause such as trauma, malignancy, endocrine disorders or antibiotic therapy. In these cases *Candida* spp, normally present as commensals in the mouth, assume the role of *opportunist pathogens*. There may be areas of white plaque or of erythematous inflammation with little plaque present. *Candida* spp produce *denture stomatitis* when trauma is caused by ill-fitting dentures and this is often associated with *angular cheilitis*.

AUMG
(**Acute ulceromembranous gingivitis: Vincent's infection**)

Bacterial aetiology

Smears from the gingival crevice in the area show a characteristic appearance of pus cells, spiral organisms and large Gram-negative rods. Until the development of good anaerobic techniques these organisms could not be grown. Morphology alone was used to classify them as *Borrelia vincenti* and *Fusobacterium fusiforme* respectively. *Fusobacterium fusiforme* has had a chequered career taxonomically and it now appears that the large cigar-shaped Gram-negative bacilli are *Leptotrichia buccalis*. Later cultural studies reported significant increases in *Bacteroides melaninogenicus*, *Vibrio* and *Campylobacter* spp in this form of gingivitis. The potential for producing necrotising and ulcerative lesions on skin appears to reside with *Bacteroides melaninogenicus* and not with the spirochaete or fusiform. Improved techniques for culturing spirochaetes are now available and this condition merits further study in the light of these technical advances.

Pathology and epidemiology

AUMG usually develops in patients with poor oral hygiene; additional stress for reasons not yet understood precipitates the development of the lesion. It is not uncommon in certain populations such as students and army recruits and it gained prominence in World War I when it was given the name of 'trench mouth'.

Laboratory diagnosis

This depends on the following findings on stained films: pus cells, Gram-negative spirochaetes and Gram-negative fusiform bacilli. Cultural methods are not satisfactory.

Treatment

Metronidazole is the drug of choice. Penicillin is also effective.

Gingivitis

Inflammation of the gingivae appears to be caused by bacteria and their products in dental plaque adjacent to the gingival margin. There is an acute inflammatory response with dilatation of gingival capillaries and exudation of fluid containing immunoglobulins (IgG), complement and polymorphonuclear leucocytes. Removal of dental plaque leads to a resolution of gingivitis. No particular organisms have been implicated in gingivitis; it is more of a reaction to the mass of organisms and the products. Actinomycetes increase in numbers as do facultative and anaerobic cocci. Some animal studies have shown that mono-infection with *Actinomyces viscosus/naeslundii* can cause gingivitis.

Periodontal (Dental) Abscess

Gram-negative anaerobic rods predominate, especially *Bacteroides gingivalis*, but a wide range of species has been reported. Anaerobic cocci, facultative streptococci and actinomycetes are also found in large numbers in samples taken from the apical region of the abscess. Perhaps significantly *Bacteroides gingivalis* is not a frequent isolate of the healthy gingival sulcus or of the early established pocket. The proteolytic activity of this organism may well contribute to its pathogenic potential.

Herpetic stomatitis

Herpes simplex virus (HSV) infection presents as primary infection or as recrudescent infection; primary infection occurs in the absence of humoral antibody. 5% of cases are clinically recognised; the patient presents with ulceration of the gums, buccal mucosa and tongue. Vesicular lesions may occur on the lips and spread to neighbouring areas of the skin; fever and local adenopathy are usually present. The incubation period is up to 1 week and virus excretion from the lesions lasts for up to 5 days. Healing takes longer—from 7 to 10 days. Type 1 is associated with such infections. Recrudescent lesions arise, despite the presence of neutralising antibody, and are found on the lips (herpes labialis) and nearby skin (herpes facialis) of chin, cheek and nose. Palatal lesions have been described, but are rare. There is no systemic upset and usually no adenopathy.

Pathology and epidemiology

Virus is usually transferred by direct contact as in kissing. Cells of the basal layer and the epithelium are infected and virus spreads to produce intra-epithelial accumulation of cell debris and exudate containing high titres of HSV. Histologically there are giant cells and intranuclear inclusion bodies in the margins of the lesion.

During the primary infection virus reaches small nerve fibres near the skin and ascends to exist as a latent infection of the neurones of the trigeminal ganglia. At intervals the virus reactivates and descends the nerve fibre. No lesion may be found, but virus can be isolated from the saliva. If infection is established, the typical vesicular lesions develop. Recrudescence is triggered by fever, another infection, menstruation, exposure of the skin to ultraviolet light and immunosuppressive treatment.

Primary infection classically occurs in young children aged 1-2 years of age but may occur at any age. Social factors and hygiene are important; patients with recrudescent lesions are a common source.

Treatment

Advice on the infections hazards to other people should be given and on the risk of auto-inoculation to fingers (herpetic whitlow) and eye and other areas of skin. Dentists and nurses, whose unprotected hands are contaminated with infected saliva, are also at risk of herpetic whitlow. Specific treatment with acyclovir ointment is available. In severe cases oral or even intravenous therapy may be necessary.

Diagnosis

Vesicular fluid can be examined in the EM and cell scrapes from the bases of lesions examined histologically and by immunofluorescence. Virus can be isolated in cell culture usually within a few days. Serological tests can show a rise in antibody titre during a primary infection: no change is detected during recrudescence.

Viral infection of salivary glands

Cytomegalovirus infection of salivary glands is clinically inapparent. The important disease is mumps.

Epidemiology and pathology

Mumps is a disease of childhood with an incubation period of about 2 weeks although it may be as long as 3 weeks. Infection is acquired by inhalation or direct inoculation, the virus localising in the salivary glands after a viraemic phase. The parotids are usually affected, but infection of the meninges, ovary and testis may occur with no salivary gland involvement. Virus is excreted via the salivary ducts and expelled in droplets. Virus excretion in saliva starts about 6 days before glandular enlargement and may persist in the urine for up to 2 weeks.

Mumps is a disease of children, although up to 20% of the adult population may be susceptible.

Diagnosis

Virus can be isolated in cell culture from saliva and urine: serological responses can also be of value.

Infection hazards of saliva

Apart from the excretion of HSV and mumps virus, CMV and EBV may be present in oral secretions. Saliva has been shown to be infectious in both acute and chronic cases of hepatitis B. There is evidence that HIV the virus associated with AIDS, is present in saliva.

FURTHER READING

Burnett G W, Schuster G S 1978 Oral microbiology and infectious disease. Student Edition Williams and Wilkins, Baltimore

Marsh P 1980 Oral microbiology. Nelson, Walton-on-Thames

Roitt I M, Lehner T 1980 Immunology of oral diseases. Blackwell Scientific Publications, Oxford

Ross P W, Holbrook W P 1984 Clinical and oral microbiology. Blackwell Scientific Publications, Oxford

Silverstone L M, Johnson N W, Hardie J M, Williams R A D 1981 Dental caries: aetiology, pathology and prevention. MacMillan, London

9

Infections of the gastrointestinal tract and liver

Normal flora

The stomach and upper small intestine in both adults and children are generally sterile in Europeans, although in certain Eastern races, due presumably to different dietary habits, streptococci and lactobacilli may be present; in general diet has a major effect on faecal flora. Once the ileum is reached, a typical Gram-negative bacillary flora is seen, composed mainly of *Escherichia coli*. The large intestine has a dense, varied flora the composition of which is influenced by the distribution and composition of the intestinal contents, local defence mechanisms, physiology of the host and diet. The organisms include *Escherichia coli*, *Streptococcus faecalis*, *Bacteroides* spp, *Lactobacillus* spp, *Clostridium* spp, anaerobic cocci, *Bifidobacterium* spp, *Klebsiella* spp, *Proteus* spp and *Pseudomonas* spp.

Defences against infection (see Table 9.1)

Table 9.1 Defences against infection in the alimentary tract

Dietary factors
Intact normal flora
Low pH of gastric juice
Non-specific mucosal factors—lactoferrin and lysozyme
Sub-mucosal histiocytes
Responses—the inflammatory response, interferons and the specific immune response (antibody and cell-mediated)

Potential pathogens (see Fig. 9.1)

The alimentary tract is an important route of infection both for agents which remain within the tract, as in gastroenteritis, or

146 CLINICAL MICROBIOLOGY

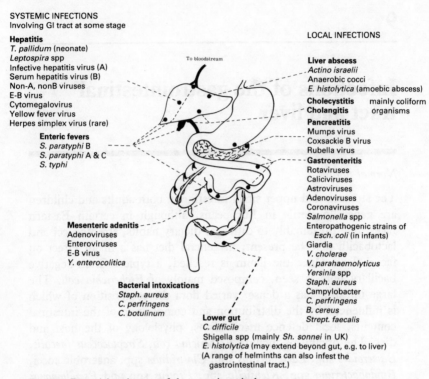

SYSTEMIC INFECTIONS
Involving GI tract at some stage

Hepatitis
T. pallidum (neonate)
Leptospira spp
Infective hepatitis virus (A)
Serum hepatitis virus (B)
Non-A, nonB viruses
E-B virus
Cytomegalovirus
Yellow fever virus
Herpes simplex virus (rare)

Enteric fevers
S. paratyphi B
S. paratyphi A & C
S. typhi

Mesenteric adenitis
Adenoviruses
Enteroviruses
E-B virus
Y. enterocolitica

Bacterial intoxications
Staph. aureus
C. perfringens
C. botulinum

To bloodstream

LOCAL INFECTIONS

Liver abscess
Actino israelii
Anaerobic cocci
E. histolytica (amoebic abscess)
Cholecystitis   mainly coliform
Cholangitis   organisms
Pancreatitis
Mumps virus
Coxsackie B virus
Rubella virus
Gastroenteritis
Rotaviruses
Caliciviruses
Astroviruses
Adenoviruses
Coronaviruses
Salmonella spp
Enteropathogenic strains of
   Esch. coli (in infants)
Giardia
V. cholerae
V. parahaemolyticus
Yersinia spp
Staph. aureus
Campylobacter
C. perfringens
B. cereus
Strept. faecalis

Lower gut
C. difficile
Shigella spp (mainly Sh. sonnei in UK)
E. histolytica (may extend beyond gut, e.g. to liver)
(A range of helminths can also infest the
gastrointestinal tract.)

Fig. 9.1 Potential pathogens of the gastrointestinal tract

which penetrate and spread to distant sites as in the enteric fevers or poliomyelitis, causing minimal effects on the gut.

The clinical features of gut infections are those of *gastroenteritis*; nausea, vomiting, diarrhoea and abdominal pain, either diffuse and/or colicky. Infections of the lower alimentary tract, such as dysentery, may cause diarrhoea predominantly and also straining and tenesmus: blood and mucus may be noticed in the stool. Vomiting is the predominant complaint in some cases, especially the intoxication associated with *Staphylococcus aureus*.

In any alimentary infection fluid and electrolyte loss can be very considerable. This is particularly so in cholera and gastroenteritis in the young; dehydration is thus the important problem.

The term *food poisoning* is used to describe infections which are usually transmitted via food contaminated at source or during preparation.

Gastroenteritis

Viral infections

Rotaviruses are the most important cause of acute non-bacterial gastroenteritis in infancy and childhood and are the most frequent causes of hospital admissions of children with acute diarrhoea. In temperate climates these viruses cause epidemics in the early months of the year, but infection can occur at any time. Infection is common in tropical regions and makes a significant contribution to infant mortality from diarrhoea.

The infection is spread by faecal contamination which occurs readily in the early stages of infection when there are upwards of $10^{10}-10^{11}$ virus particles per gram of stool, and the possibility of respiratory spread is suggested by the frequent occurrence of respiratory symptoms. Symptoms start within 2–3 days and last for a similar time although virus excretion persists for up to 1 week in most cases. In residential homes and nurseries outbreaks can be explosive. Infection in adults is usually acquired from children and is mild, but may be more severe in the elderly, with mortality rates up to 10% reported. Four subtypes of the virus are known.

Laboratory diagnosis

This can be made by examination of a fresh stool specimen. It can be examined in the EM, or most often by an ELISA for rotavirus antigen. There is sufficient RNA in the stool to allow detection of the characteristic 10 fragments of RNA: variation in the mobility of the RNA fragments can be used to type viruses for epidemiological studies.

Treatment

Symptomatic: vaccines are under development and some clinical trials have been reported.
Caliciviruses. These small viruses are another important cause of viral gastroenteritis. The features of the disease are similar to those caused by rotaviruses, although infection occurs most often in older children and adults. There are many of these viruses although no formal subdivision has been proposed. The original isolate was the Norwalk agent from an outbreak in Ohio in 1967. Polluted shellfish

have been implicated in several outbreaks. Infection is also prevalent in tropical countries.

Laboratory diagnosis. The only test available is EM of stool; the principal features of the viruses are the small size (25-30 nm) and lack of capsid structure.

Astroviruses. Recognised in the EM by their small size and their morphology, these viruses have been detected in individuals with diarrhoea and in some outbreaks of gastroenteritis.

Adenoviruses. Numerous typical adenovirus particles can be seen in some outbreaks of gastroenteritis: these appear to be different from the typical respiratory adenoviruses. Diagnosis is by EM as they do not grow readily in cell culture.

Coronaviruses have been seen by EM, but their role is uncertain.

Food poisoning

Bacterial aetiology

In the United Kingdom *Salmonella* spp (particularly *typhimurium*) and *Campylobacter* spp account for most cases. Other organisms implicated, in order of frequency, are *Clostridium perfringens, Staphylococcus aureus, Escherichia coli* and *Bacillus cereus*. Rarely *Vibrio parahaemolyticus* and *Streptococcus faecalis* may be involved.

Pathogenesis and epidemiology

This is an acute form of gastroenteritis due to bacterial contamination of food or drink. There are two main types (1) The *infective type* and (2) the *toxic type.*

(1) *Infective type*: Ingestion of food contaminated with salmonellae leads to this type of food poisoning. The infecting organisms multiply in the food and later in the intestine of the victim. Many different species of *Salmonella* may be responsible for this type of infection, including *Salmonella typhimurium, heidelberg, enteritidis, dublin* and *thompson*. The 12-48 hour incubation period is longer than in toxic food poisoning and is followed by pyrexia and diarrhoea of one or two days duration. Vomiting is not usually severe. Cold meats, poultry, sausages and occasionally eggs are common sources of salmonellae. Infection may be direct from an infected carcase or acquired from cases and carriers among food handlers if there is general lack of hygiene, inadequate cooking of foods and inadequate thawing of foods, particularly poultry, before cooking.

(2) *Toxic type*: This is most commonly produced by *Staphylococcus aureus* and *Clostridium perfringens*, both of which produce an enterotoxin. Staphylococcal food poisoning causes nausea, vomiting and often prostration with some diarrhoea within 1–6 hours of ingesting the contaminated food containing the pre-formed heat-stable toxin. Foods commonly incriminated include cream cakes, trifles, meat paste, pies and ham. The usual sources of staphylococci are the upper respiratory tract, particularly the nose, and septic lesions on the skin of food handlers. In *Clostridium perfringens* food poisoning, abdominal cramps followed by diarrhoea occur around 8–12 hours after ingestion of contaminated food. Fever and vomiting are rare. *Clostridium perfringens* is a sporing organism that can withstand the heat of cooking and multiply freely in conditions where food is allowed to cool slowly or where it is gently reheated before serving. Foods commonly associated are bulk meat dishes such as stews. The source of organisms may be either the meat itself or the intestine of the food handler. A dose of 10^7–10^8 organisms is required to initiate infection and the symptoms of infection are due to an enterotoxin which is produced in the intestine, and not preformed in the food. Enterotoxin production is associated with sporulation of the organisms. In general, food poisoning is far more common than it should be and is related to the many commercially-prepared foodstuffs and catered meals. The summer is the commonest season for outbreaks because of high ambient temperatures and lack of refrigeration of prepared foods, allowing multiplication of organisms.

Laboratory diagnosis

Specimens of vomit, faeces and suspected foods should be sent for analysis and bacterial culture. Salmonellae when identified may be phage-typed and *Clostridium perfringens* isolates may be serotyped.

Prophylaxis

Meat inspection, routine medical inspection of food handlers and education in matters of hygiene for all food handlers should be encouraged. Hygienic premises, adequate cooking and use of refrigeration are essential. Raw and cooked meats must be kept separately.

Treatment

Fluid balance, particularly in the elderly and the very young, must be maintained. The use of antibiotics is not recommended in most cases as they do not shorten the duration of the diarrhoea and do not eliminate the causal organisms.

The following are less common causes of food poisoning in the United Kingdom:

Bacillus cereus. This is associated with rice, and symptoms are probably due to preformed enterotoxin. Two clinical types are described: (1) staphylococcal-like, of rapid onset and with vomiting as the main feature, and (2) clostridial-like, with a longer incubation period and with diarrhoea and abdominal pain.

Vibrio parahaemolyticus. The organism producing this type of food poisoning is a marine vibrio associated with shellfish. The organism multiplies in the intestine and may produce a cholera-like enterotoxin. Patients usually suffer from vomiting, diarrhoea and abdominal pain.

Enteric fever

Bacterial aetiology

The term *enteric fever* includes typhoid and paratyphoid A, B and C infections caused by *Salmonella typhi* and *Salmonella paratyphi* respectively. Other serotypes of *Salmonella* do not cause enteric fever but *salmonellosis*, a form of food poisoning.

Pathogenesis and epidemiology

Figure 9.2 shows the pathogenesis of enteric fever and it can be seen that bacteriaemia is the important feature of the illness; the bacilli are also present in large numbers in many organs. The pathogenicity of salmonellae appears to depend both on their ability to survive and grow inside macrophages and on the toxicity of their lipopolysaccharide *O antigen*. In addition, typhoid bacilli possess a glycolipid microcapsule called the *virulence* (*Vi*) antigen which protects the organism against phagocytosis and antibody. Infection often presents as a pyrexia of unknown origin (p.u.o.).

The source of typhoid and paratyphoid infections is *man*, either as a case or as a carrier and spread can be via water, food, or the faecal-oral routes. The most common cause of enteric fever in the United Kingdom is *Salmonella paratyphi B*.

INFECTIONS OF THE GASTROINTESTINAL TRACT

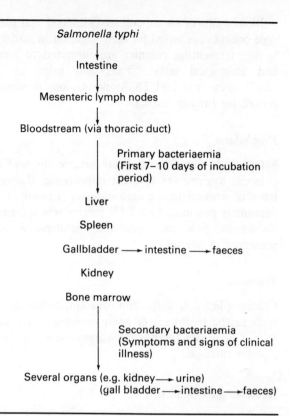

Fig. 9.2 Pathogenesis of enteric fever

After an attack of enteric fever the organisms may persist in the gall bladder and are excreted intermittently in the faeces.

Laboratory diagnosis

Specimens required for diagnosis vary according to the stage of the infection. During the first 7–10 days blood culture is the examination of choice and in the second and third weeks stool cultures are usually positive; urine cultures may also be positive at this stage. A test for agglutinating antibody in serum, the Widal reaction, can be done in the second or third weeks but is of limited value as false positive results are common. Blood, stool and urine cultures are the most important tests and should be done repeatedly.

Blood cultures are subcultured on blood agar plates and coliform-type colonies are tested for their sugar fermentation reactions. Non-lactose fermenting colonies are subjected to further biochemical and serological tests. Stools and urine are subcultured on MacConkey agar and DCA; non-lactose fermenting colonies are picked for further testing.

Prophylaxis

Safer water supplies, adequate sewage disposal and attention to personal hygiene are of great importance. Bacteriological surveillance of workers in the food industry is essential. Immunisation is commonly practised with TAB vaccine which consists of heat-killed *Salmonella typhi* and *Salmonella paratyphi* A and B organisms preserved in phenol.

Treatment

Chloramphenicol, ampicillin and cotrimoxazole are useful drugs both in the treatment of acute typhoid fever, and of the carrier state. Cholecystectomy is often required when there is chronic gall bladder carriage.

Infantile gastroenteritis

Bacterial aetiology

Enteropathogenic strains of *Escherichia coli* that are responsible for infantile gastroenteritis are limited to a small number of 'O' serotypes with an associated B-type K antigen, for example 026,B6; 055,B5 and 0111,B4.

Pathogenesis and epidemiology

An enterotoxin similar to cholera exotoxin in its mode of action is produced; some strains may be locally invasive like shigellae or penetrate more deeply like salmonellae. In infants the disease is sometimes known as *cholera infantum* and causes a massive loss of fluid from the small intestine, often resulting in peripheral circulatory failure.

Outbreaks occur in infant nurseries and day nurseries as well as in the normal population and most patients are under 18 months of age. There is a higher incidence in artificially-fed babies.

Immunity in older children and adults may be due to acquired local immunity in the intestine.

Laboratory diagnosis

Faeces are plated on blood agar and MacConkey agar. Agglutination tests are performed on *Escherichia coli* isolated using sera containing antibodies to O and B antigens.

Prophylaxis

Attention to personal and environmental hygiene, proper nutrition and breast feeding of babies will lessen the incidence of the disease.

Treatment

Rehydration and restoration of the electrolyte balance is vital. Antibiotic chemotherapy may be of benefit in mild cases but cannot act quickly enough in the acute stages.

Campylobacter enterocolitis

Bacterial aetiology

Campylobacter coli and *jejuni*.

Pathogenesis and epidemiology

Campylobacter organisms are significant causes of human diarrhoea. The habitat of these organisms is the gut of farm animals, such as cattle, sheep and pigs and poultry, and pets, notably puppies. Poultry are important sources, indeed campylobacter organisms can be isolated from about 75% of broiler chicks after slaughter. If cattle are carriers man becomes infected by drinking contaminated milk.

Campylobacter enteritis and colitis, which comprises fever, abdominal pain, malaise and produces frequent and offensive watery, bloody stools is self limiting after a few days. Outbreaks of infection are more common in late summer.

Laboratory diagnosis

Organisms are Gram-negative curved or spiral rods which are

highly motile. They are grown in 7% oxygen, 10–15% carbon dioxide mixed with either hydrogen or nitrogen. The optimal growth temperature is 43°C and the organisms, which look like droplets on media, grow within 24–48 hours. Selective media containing antibiotics are used if faeces are under investigation.

Treatment

Erythromycin and aminoglycosides are useful drugs.

Dysentery

Bacterial aetiology

The genus *Shigella* contains four groups named *Shigella dysenteriae, flexneri, boydii* and *sonnei*. Each group is subdivided into various serotypes except *Shigella sonnei* which is serologically homogeneous. In addition amoebae and *Giardia lamblia* may be involved.

Pathogenesis and epidemiology

Bacillary dysentery occurs by ingestion of the organisms. The bacilli attach themselves to the epithelial cells of the mucosal villi, enter these cells, multiply within them and spread into adjacent cells. The infected cells are killed and an inflammatory reaction results in the submucosa and lamina propria with resulting necrosis and ulceration of the epithelium. Invasion of the deeper layers does not generally take place. Stools contain blood, pus and mucus. *Shigella dysenteriae* infection is usually imported and causes a particularly severe form of dysentery because of the production of a powerful enterotoxin, in addition to the lipopolysaccharide endotoxin formed by all *Shigella* species. *Shigella flexneri* and *Shigella boydii* cause a less severe illness and *Shigella sonnei* infection which is the most common in the United Kingdom may only produce a few loose stools with slight abdominal discomfort.

In the main bacillary dysentery is spread from hand to mouth. A case or carrier can contaminate his hands at toilet and further contaminate door handles, lavatory chains and hand towels if his personal hygiene is deficient. Subsequent handling of these by another person results in the transfer of the dysentery bacilli to the hands and possibly the mouth. Pre-school and primary school children are frequently involved in outbreaks of dysentery. Occasionally outbreaks may be food- or water-borne.

Laboratory diagnosis

A stool sample should be examined; this is much more satisfactory than a rectal swab. The swab is plated on MacConkey media and selective media such as deoxycholate citrate agar (DCA). (Xylose lysine deoxycholate agar (XLD) may be used as an alternative to DCA.) A fluid medium such as selenite broth should also be inoculated. Non-lactose fermenting colonies are then subjected to biochemical tests, and identified serologically.

Prophylaxis

Attention to hygiene is very important, for example the use of paper towels or hot air blowers rather than roller towels in institutions and the installation of washbasin taps operated by foot instead of by hand. Prophylactic use of antibiotics is a worthless exercise.

Treatment

Antibiotics have little part to play in this disease and may indeed prolong the period of excretion of the bacilli. An exception is the treatment of *Shigella dysenteriae* infection for which cotrimoxazole, ampicillin and tetracyclines may be used.

Amoebic dysentery

Entamoeba histolytica invades the mucosa of the colon and can cause moderate or fulminating diarrhoea. This is predominately a disease of the tropics. Metronidazole is the drug of choice.

Cholera

Bacterial aetiology

Two vibrios, the classical *Vibrio cholerae* and the *El Tor* biotype cause cholera but other vibrios such as *Vibrio parahaemolyticus* can cause an acute form of food poisoning. The two pathogenic biotypes can each be divided into two serological subtypes named *Inaba* and *Ogawa*.

Pathogenesis and epidemiology

Vibrios are ingested in food and water and surviving the acid barrier of the gastric juice begin to multiply in the intestinal contents, becoming attached to the epithelial cells of the small intestine. As they multiply in the lumen they produce a powerful enterotoxin whose action promotes increased outpouring of water and electrolytes into the intestine. There is no inflammatory reaction of the mucosa nor invasion of the intestinal wall. Lymph glands and blood are not affected. Effects of the disease such as severe dehydration are due to the massive outpouring of water and electrolytes. The resulting 'rice water' stools contain much mucus.

Cholera is highly communicable, transmitted from person to person via contaminated water or uncooked foods. During non-epidemic periods in endemic areas there is a high ratio of symptomless carriers to clinical cases particularly with the *El Tor* biotype, although explosive epidemics occur if a vehicle such as food or water is infected.

Laboratory diagnosis

A specimen of stool is examined for vibrios. In urgent cases a *vibrio immobilisation test* can be done under dark-field microscopy using specific antisera but it is most important to culture the organism by inoculation onto selective media such as bile-salt containing agar, and into a tube of alkaline peptone to encourage growth of vibrios should their numbers be small. Precise identification including biotyping and serotyping depends on further biochemical, serological and phage-sensitivity tests.

Prophylaxis

This includes isolation, treatment of infectious cases, surveillance of contacts and enforcement of proper community and group standards of hygiene. Prophylactic immunisation is not recommended because the available vaccines produce immunity that does not last longer than a few months.

Treatment

The most important treatment is to restore the vast loss of fluids and electrolytes. Later tetracyclines may be given to shorten the illness and reduce fluid requirements.

Travellers' diarrhoea

Contaminated food or water are generally responsible and although many bacteria can cause this, the important organisms are enterotoxigenic strains of *Escherichia coli*. Protozoal infections such as amoebiasis, giardiasis and balantidiasis must also be borne in mind as well as helminth infections.

Yersinia diarrhoea

This acute infection of the terminal ileum, which can mimic appendicitis, occurs mostly in children; the causative organisms are *Yersinia enterocolitica*, the sources of which are wild and domestic animals. Drinking contaminated water is the usual means of infection. Chloramphenicol or cotrimoxazole are used in treatment.

Pseudomembranous colitis (PMC)

Clostridium difficile is carried quite commonly in the faeces of healthy persons. It has been associated with PMC as a complication of treatment with antibiotics such as ampicillin, but particularly the lincomycins. Its role in this antibiotic-associated colitis which can often be severe is unclear, though it is known to produce enterotoxin and cytotoxin. Whether antibiotic treatment encourages seeding of the gut with these clostridia or whether they are selected out by the antibiotic and then multiply in the gut is not fully established. There is evidence of spread of the organisms from patient to patient in hospital wards. Oral vancomycin is used after discontinuation of any previous antibiotic treatment.

Giardiasis

Caused by a flagellate protozoon *Giardia lamblia* which is a common bowel pathogen throughout the world, including the United Kingdom. Flatulence and diarrhoea are produced. Metronidazole is used in treatment.

Brucellosis

Bacterial aetiology

Three main species of *Brucella* occur in different animal hosts:
Brucella abortus: Cattle
Brucella melitensis: Goats and sheep
Brucella suis: Pigs

Pathogenesis and epidemiology

Each of the three species is pathogenic to man. *Brucella abortus* is the most common in the United Kingdom. The organisms produce contagious abortion in cattle and then localise in the mammary glands. Uterine discharges, urine, faeces and milk may all contain the bacilli and human infections occur by handling infected carcases, by contact with infected discharges, or by drinking infected milk or milk products. Infection can be contracted through abrasions in the skin and through the alimentary and conjunctival mucosae. Human brucellosis occurs predominantly in farmers, veterinary workers and in those who drink unpasteurised milk but as it is a zoonosis there is no man-to-man spread.

The bacteria localise and multiply inside the cells of the reticulo-endothelial system forming granulomatous lesions in the liver and spleen where they are inaccessible to the circulating phagocytes. Clinically the disease may be acute, subacute or chronic. Subclinical infections also occur and these are common in veterinary surgeons.

Laboratory diagnosis

Repeated *blood cultures* should be taken, although they are often negative in the disease, and they should be incubated in an atmosphere of 5–10% carbon dioxide for 3–4 weeks before being discarded. Castenada's method using a bottle that contains solid and fluid media may be useful. After isolation species differentiation depends on biochemical tests.

Serological tests are helpful in the second week of clinical infection, as both agglutinating (IgM) antibodies and complement-fixing (IgG) antibodies are present in the acute stage. In chronic brucellosis IgM antibodies decrease and may be absent even when the patient is clinically ill.

Antiglobulin tests (Coombs' test) can also be done.

Prophylaxis

Animals must be tested and positive reactors slaughtered or segregated. It is possible to vaccinate animals, for example female calves, against the disease but because of undesirable side effects this is not recommended for man. A milk ring test (MRT) is used to screen for infection in dairy cattle. Pasteurisation of milk has also helped reduce the incidence of the disease.

Treatment

Many antibiotics are effective in acute infections, particularly a combination of oral tetracycline and intramuscular streptomycin. Chemotherapy is ineffective in the treatment of chronic infection.

Abscesses

Pyogenic liver abscess

The most common route of infection is via the hepatic portal vein, often as a sequel to diverticulitis or abdominal, pelvic and perianal surgery. Other routes are the hepatic artery and the biliary tree, or by direct extension.

Streptococcus milleri are the organisms most commonly isolated, either separately or in combination with anaerobic bacilli. Treatment is by drainage with or without antibiotic cover. *Entamoeba histolytica* may also cause liver abscess. *C. trachomatis* causes a perihepatitis.

Subphrenic abscess

Bacterial aetiology

Staphylococcus aureus and, in particular, organisms found in the intestinal flora such as coliform organisms and *Streptococcus faecalis*, are involved. The flora is commonly mixed.

Pathogenesis

The peritoneum in this area is arranged in a complicated fashion with the result that there are seven spaces in which pus may collect, three on either side of the body and one in the midline. A subphrenic abscess can be caused as a sequel to a perforated peptic ulcer, trauma, appendicitis and leakage after gall bladder and stomach operations. Empyema may be associated.

Laboratory diagnosis

Aspiration of the pus is the best way to obtain a specimen for bacterial diagnosis; pus is Gram-stained and cultured aerobically and anaerobically. Blood cultures should be taken.

Treatment

This is surgical drainage with or without antibiotic therapy.

Pelvic abscess

Bacterial aetiology

This is much the same as for subphrenic abscess. Organisms from the faecal or vaginal flora including anaerobes are commonly involved.

Pathogenesis

This can be a complication of appendicitis or salpingitis, and there may be an associated peritonitis.

Laboratory diagnosis and treatment

As for subphrenic abscess; surgical drainage is essential.

Peritonitis

Bacterial aetiology

This is variable depending on the source of infection and includes normal intestinal bacteria such as coliform organisms, *Streptococcus faecalis*, anaerobic cocci and *Bacteroides* spp. Frequently more than one organism is isolated.

Pathogenesis

Infection can be *direct* or *septicaemic*. *Direct* infection originates from perforation of an area of the intestine from trauma or following surgery. The appendix, gall bladder, fallopian tubes and damaged intestine, for example from a carcinoma, are common sources. Peritonitis may remain *localised* when the cause is an inflamed appendix or become *generalised* when caused by perforation of a peptic ulcer.

Laboratory diagnosis

Any exudate or pus that can be obtained from drainage should be examined microscopically and cultured aerobically and anaerobically. Bacteriaemia frequently occurs in conjunction with peritonitis and blood cultures must be taken.

Treatment

Surgical treatment is frequently essential and antibiotics, often in combinations, are urgently required. If treatment has to be blind the possibility of the involvement of anaerobic organisms must be considered. A penicillin or clindamycin may be administered along with an aminoglycoside such as gentamicin. Metronidazole is effective against *Bacteroides* spp.

Bacteria and malignant disease

The nature of the diet affects the composition of the intestinal flora and determines the substrates available for bacterial metabolism and a relationship between diet and malignant disease of the intestine has been postulated. The incidence of malignant disease, in particular cancer of the colon, has been shown to correlate strongly with the amount of dietary fat and animal protein and not at all with dietary fibre. It is thought that carcinogenic substances may be produced by the action of bacteria on certain dietary components and intestinal secretions.

INFECTIONS OF THE LIVER

Hepatitis

Viral infection of the liver presents as malaise, loss of appetite and fever: jaundice and abnormal levels of amino transferase enzymes indicate hepatocyte injury. Hepatitis occurs as part of a disseminated infection, as in rubella, CMV, HSV and enteroviral infections of the newborn. In older patients, hepatitis is almost universal in patients with glandular fever, and may be due to CMV infection especially in the immunocompromised patient. In tropical Africa and South America yellow fever is important, and there are many other causes, including chronic infection with *Coxiella burneti*. However, viral hepatitis usually implies infection with the virus of hepatitis A, hepatitis B (and the delta agent) and non-A non-B hepatitis.

Hepatitis A or infectious hepatitis

This is caused by a small RNA virus with the features of a picornavirus. Infection is spread by the faecal-oral route: virus is excreted in faeces late in the incubation period, of about 1–3 weeks, and only for a day or two after the onset of jaundice. Mortality is rare and occurs in less than 1 in 1000 cases from fulminant hepatic failure. No carrier state is known and there are no chronic sequelae of infection. Only 1 virus type is recognised: immunity is life long after an attack.

Epidemiology

The virus is world-wide: most infections occur in the young with a very low clinical attack rate. In the developed countries infection may be interrupted due to good standards of hygiene and sewage disposal. If infection is delayed to older ages the clinical attack rate is increased significantly. Food and water-borne outbreaks occur, and shellfish are frequently incriminated as concentrating the virus from sewage discharged into the sea. Cooking of the contaminated shellfish is seldom adequate to inactivate the virus.

Laboratory diagnosis

Virus can be detected in faecal samples by electronmicroscopy, and the virus has been grown on fetal rhesus monkey-cell cultures. However there is little chance of virus detection by the time the patient seeks advice. In these circumstances diagnosis is made by the examination of a serum sample for antibody to the virus. The serum titres are usually stable in acute and convalescent sera. Thus diagnosis depends on the detection of virus-specific IgM antibody in the acute phase sample.

Control

Transmission of infection has usually occurred before clinical features appear, therefore very few precautions are necessary other than safe disposal of faeces during the acute phase if the patient is in hospital. Passive immunisation with pooled immunoglobulin is advised for persons travelling and working in endemic areas. If the stay is prolonged for more than 6 months, a further dose must be administered. No active immunisation is available, although development should be possible with the availability of cell cultures for the virus.

Hepatitis B or serum hepatitis

This has an incubation period of more than 40 days (range 60–90 days) and may be as long as 6 months; it is dependent on the dose of challenge virus. Clinically the symptoms are similar to those of infectious hepatitis, although rash and arthralgia occur in the prodromal stage before the onset of jaundice. Most patients recover, but 5–10% of cases become long-term carriers. The mortality rate is considered to be about 1% overall, although higher rates have been recorded.

Chronic sequelae. Many carriers remain asymptomatic although some develop progressive liver disease and cirrhosis. Long term carriage is an important predisposing factor in the development of hepatocellular carcinoma, a major tumour in parts of tropical Africa and the Far East.

Epidemiology

Transmission of infection was first established via blood transfusion and blood products, but semen and saliva are also infectious, and must account for the transmission of most HBV infections in the world. Parenteral injection and mucous membrane contamination are effective means of transmission. The carriage rates are 0.1% in the United Kingdom to 5–15% in the tropics and the Far East. In areas of high endemicity infection is acquired from an infectious mother at birth, from other infected children or even *in utero*. In the United States and United Kingdom, carriage rates are highest in immigrant groups, parenteral drug abusers and the promiscuous, especially male homosexuals. Rates of 5–10% are often found in these groups: the risk of transmission to health staff arises when such patients need care. The risks are greatest when blood contact occurs, especially if this is uncontrolled and needles and sharps are in use. Thus those who work in surgical and obstetrics units are most at risk, although dentists and laboratory workers can also be exposed to infection. Within hospitals, risk has been associated with renal dialysis units, although this has been controlled in the United Kingdom by screening and measures to limit the spread of infection. Patients and staff in hospitals for the mentally-handicapped are often at risk and carriage rates of 5–10% are common, especially in the Down's Syndrome group. Despite routine screening of blood donations post transfusion hepatitis B can still occur: the risk is greater in recipients of certain blood products made from large numbers of donors. Haemophiliacs almost always show a high prevalence of infection. Patients whose immune responses are suppressed are more at risk of becoming carriers, as are the very young.

Diagnosis

A sample of clotted blood should be sent to the laboratory: the sample must be in a properly sealed container, separate from the request form. The plastic bag carrying the specimen should be marked 'Hepatitis Risk' *or* 'Risk of Infection.'

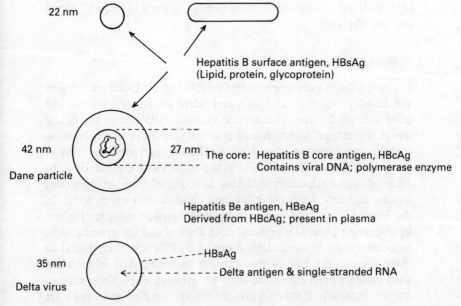

Fig. 9.3 The particles and antigens of the hepatitis B and delta viruses

A variety of tests is available for the diagnosis of hepatitis B in both acute and chronic stages. The components of the virus and the antigens used in diagnostic tests are shown in Fig. 9.3.

Detection of HBsAg is the mainstay of diagnosis in both acute and chronic cases. These states can be differentiated by the presence of IgM anti-HBc. A further antigen, the hepatitis Be antigen (HBeAg), can be detected in the plasma. Its presence correlates well with the production of Dane particles and hence with infectivity. It is not associated with the particles and is derived from the HBcAg.

The sequence of events represented in Fig. 9.4 indicates the main events during an acute infection: it must be emphasised that the timing and duration of HBsAg, HBeAg and appearance of anti-HBs are variable.

Carriers initially are HBeAg positive for more than 6 months. Thereafter over a period of years to decades they may become less infectious as HBeAg is lost and anti-HBe appears. This sequence is comparable to the acute stage events in Fig. 9.4, but with a much prolonged time scale there may be a period of increased liver damage. Integration of viral DNA into host cell chromosomes may take place at this late stage.

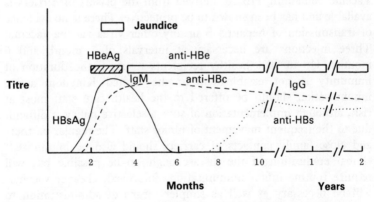

Fig. 9.4 Sequence of events in a case of acute hepatitis B followed by recovery

Control

No special precautions are necessary in patient care unless the patient is bleeding or requires full nursing procedures. Contamination of unprotected skin and mucous membranes with blood and secretions must be avoided. The use and disposal of needles and instruments requires care. Simple hygienic measures, for example hand washing and wearing disposable gloves, are of great importance. Patients must be advised of the risk of transmission to their close family contacts: the use of barrier contraception methods is recommended.

Post-exposure therapy

If a needle-stick injury, cut or mucous membrane contact occurs with blood from a patient who is HBsAg positive and especially also if HBeAg positive hyperimmune hepatitis B immunoglobulin (HBIG) should be given as soon as possible, preferably within 24–48 hours of the incident. The dose may be repeated 1 month later. Active immunisation should also be offered; a combination of immediate passive and active immunisation must be given to babies born to infected mothers to reduce the considerable risk of the baby becoming a carrier. Clinical hepatitis is rare in such infants.

Active immunisation

Vaccine containing HBsAg derived from the plasma of carriers is available and has been shown to be protective. There is no evidence of transmission of hepatitis B or any other virus via the vaccine. Three injections are necessary at intervals of 1 month and 6 months. Up to 90% healthy recipients respond: the duration of immunity is not established. In the United Kingdom active immunisation should be offered to the health care staff most at risk, although implementation of such a selective policy is difficult due to the frequent movement of junior staff. The regular partners and close family contacts of carriers should also be immunised. Global eradication of the disease ought to be possible but will require routine infant immunisation. Improved, cheaper vaccines will be necessary as well as simpler means of administration to achieve this worthwhile objective.

Non-A, non-B hepatitis

With the advent of reliable and sensitive tests for hepatitis A and B it became apparent that a varying proportion of clinical hepatitis could still not be diagnosed. Some are due to other viral infections such as CMV and EBV viruses, and others may be due to drugs. However, non-A, non-B viruses are estimated to cause from 5 to more than 50% of cases of jaundice in hospital. The majority (90%) of post-transfusion hepatitis is of this type. There is evidence that there are several different agents. Large outbreaks have been recorded in the East, with epidemiological features similar to hepatitis A. Most attention has focussed on the hepatitis B-like infections acquired from blood and blood products, thus the same patients are at risk as described for hepatitis B. Clinically the degree of hepatitis is less than with the B virus, although chronic infection is more common: epidemiological studies indicate that there are several viruses. There is no definitive laboratory test; transmission studies in chimpanzees indicate that one of the agents is sensitive to ether and is associated with the development of microtubules within infected cells.

FURTHER READING

Beasley R P, Hwang L-Y, Lin C-C, Chien C-S 1981 Hepatocellular carcinoma and hepatitis B virus. Lancet ii: 1129–1133

Beasley R P, Hwang L-Y, Lee G C-Y et al 1983 Prevention of perinatally transmitted hepatitis B virus infection with hepatitis B immune globulin and hepatitis B vaccine. Lancet ii: 1099–1102

Christie A B 1980 Infectious diseases: epidemiology and clinical practice, 3rd edn. Churchill Livingstone, Edinburgh

Collee J G 1974 Bacterial challenges in food. Postgraduate Medical Journal 50: 636–43

Colombo M, Manancci P M, Carwells V et al 1985 Transmission of non-A, non-B hepatitis by heat-treated factor VIII concentrate. Lancet ii: 1–4

Geddes A M 1973 Enteric fever, salmonellosis and food poisoning. British Medical Journal 1: 98–100

Hendrickse R G 1972 Dysentery including amoebiasis. British Medical Journal 1: 669–672

Lowbury E J L, Ayliffe G A J, Geddes A M, Williams J O 1975 Control of hospital infection: a practical handbook. Chapman and Hall, London

Meers P D 1981 Campylobacter enteritis and the community. British Medical Journal 286: 243–244

Rizzetto M 1983 The delta agent. Hepatology 3: 719–737

Tedder R S 1983 Towards the control of hepatitis B. In: Waterson A P (ed) Recent Advances in Clinical Virology: 217–236

Welsby P D, Smith C C 1978 Brucellosis: the current position. British Journal of Hospital Medicine 19: 20–27

Zuckerman A J 1983 Vaccines against hepatitis B. In: Waterson A P (ed) Recent advances in clinical virology. Churchill Livingstone, Edinburgh, pp 205–216

10

Infections of the urinary tract

Normal flora of the urethra

The renal tissue, ureters bladder and proximal urethra are sterile but several species of organisms may be found as commensals in the distal urethra and most of these are derived from faecal flora. Organisms include *Escherichia coli*, *Proteus* spp, *Klebsiella* spp, *Pseudomonas* spp, various other coliform organisms, staphylococci, streptococci, lactobacilli, mycoplasmas and ureaplasmas.

Defences against infection

Bladder urine is normally sterile because of hydrokinetic and mucosal factors. Hydrokinetic aspects include the periodic voiding of urine, the constant dilution of residual urine in the bladder by inflow from the kidneys and the completeness of emptying of the bladder. The bactericidal mechanisms in the bladder mucosa cause a rapid clearing of bacteria from the mucosa, the mechanism of which is unclear but may be immunological. Local antibody, complement and lysozyme may also be protective. Prostatic secretions and periurethral gland secretions also possess antibacterial factors.

Potential pathogens (see Table 10.1)

Urinary tract infection (UTI)

Bacterial aetiology

Escherichia coli is responsible for the greatest number of cases but many other bacteria are involved including *Streptococcus faecalis*, *Proteus* spp, *Pseudomonas aeruginosa*, *Klebsiella* spp, *Staphylococcus epidermidis* (*albus*) and *saprophyticus*.

Table 10.1 Potential pathogens of the urinary tract

Infection	Organisms
Pyelonephritis	Esch. coli Proteus Pseudomonas Strept. faecalis Staph. epidermidis Staph. saprophyticus Klebs. aerogenes
Cystitis	Myco. tuberculosis C. trachomatis
Urethritis	N. gonorrhoeae Ureaplasma urealyticum C. trachomatis Trichomonas
Prostatitis	Coliforms N. gonorrhoeae C. trachomatis

Pathogenesis and epidemiology

Although terms such as cystitis and pyelonephritis are used, UTI is not always confined to one particular anatomical site and the urinary tract should therefore be considered as a whole. Attempts to localise UTI clinically may be difficult.

Frequency and *dysuria* are common symptoms, especially in females, of all ages, but a positive urine culture is found in only a proportion of cases. Culture-negative and culture-positive cases are clinically indistinguishable. Infection generally spreads directly from the perineum, aided in females by the short distance between urethra and anus, the interposition of the vaginal introitus and the shortness of the urethra. Haematogenous spread is rare. Various factors contribute to this type of infection such as poor personal hygiene, the use of tampons, sexual intercourse, pregnancy and, in older women, gynaecological problems such as prolapse and vaginitis. Incomplete emptying of the bladder due to mechanical obstruction or neurological defects may also produce a source of residual urine that can become infected.

Asymptomatic bacteriuria

Transient symptomless bacteriuria is common in the female and is usually a benign condition with spontaneous resolution, although in the pregnant woman or in childhood there can be progressive renal disease and damage. If there are congenital abnormalities in the urinary tract or some form of obstruction bacteriuria may persist and produce symptoms.

Renal infection may follow infection of an obstructed urinary tract, for example due to the presence of calculi. Vesicoureteric reflux is reasonably common particularly in children and is of paramount importance in the production of renal damage; the reflux can carry organisms into the substance of the kidney and it is also a source of residual urine in the bladder. In addition it may produce damage by back pressure, resulting ultimately in chronic pyelonephritis; renal scarring is not common after infection in the absence of reflux. The renal medulla is very susceptible to infections; it is the site of ammonia production which has a powerful anti-complementary effect.

Laboratory diagnosis

Bacteriological examination of the urine is essential for confirmation of the diagnosis, control of chemotherapy and assessment of the results of treatment.

The diagnosis of urinary tract infection relies principally on the demonstration of significant bacteriuria by quantitative culture of a fresh or refrigerated midstream specimen of urine (MSSU). A viable count of above 10^5 organisms of a single bacterial species per ml of urine (10^8 per litre) indicates a *true infection* (significant bacteriuria); between 10^4 and 10^5 organisms per ml is of *doubtful significance* and may indicate either infection or heavy contamination (only a few specimens fall into this 'doubtful' group); 10^4 organisms per ml are sometimes found in the absence of infection and also when a patient is on antibiotic therapy.

Specimen collection

Proper collection of specimens is essential and because urine is a good culture medium it is important that specimens reach the laboratory with the minimum of delay, or that they are refrigerated until despatched. The following specimens can be obtained:

Midstream specimen (MSSU) (see Ch. 4).

Suprapubic aspiration of bladder urine. If it is difficult to collect an MSSU, or previous findings have been equivocal, it may be worthwhile to consider suprapubic aspiration.

Urine specimens from infants. Bag specimens are unsatisfactory. A 'clean catch' specimen is adequate if the infant can be caught in the act, but suprapubic aspiration is a simple procedure and gives the most satisfactory and reliable results.

Catheter specimens of urine should be submitted only if the catheter is already *in situ*. The sample should be collected directly from the catheter and not from the collection bag.

Examination of MSSU

Microscopy. This is useful for detecting pus cells, epithelial cells, red blood cells and bacteria and should be performed on a *centrifuged* deposit. Both wet films and Gram-stained films may be examined.

Culture. Specimens are usually cultured on nutrient agar and MacConkey agar.

Quantitative culture. This is done using a well-mixed specimen by the method of Miles and Misra. It is a surface viable count, and gives accurate results, but as it is a time-consuming test it cannot be used routinely in a busy service laboratory.

Semiquantitative culture. This kind of culture provides useful information. There are three commonly-used methods: (1) transfer of a fixed volume of undiluted urine to a suitable culture medium by a calibrated loop; (2) the blotting-paper strip method; (3) dipslides.

The blotting-paper strip, of known area, is dipped in the urine then impressed on MacConkey agar. Colony counts within the area can then be related to numbers of organisms per ml.

Dip slides. (Plastic slides coated with CLED agar (Cystine lactose electrolyte-deficient) on one side and MacConkey agar on the other) can be obtained commercially. A slide is dipped into the freshly-passed urine; a semi-quantitative assessment of the viable count is obtained by comparing the density of growth on the media, after incubation, with the manufacturers' graded charts. Dip-slides avoid the problems of transporting specimens, particularly from General Practitioners' surgeries, and are useful in screening for significant bacteriuria in busy out-patient clinics, for example ante-natal clinics. The incubated slides should be examined in the bacteriology laboratory. (Interpretation of mixed cultures is difficult with dip slides).

Results of semi-quantitative culture from fresh or refrigerated urine

The results may be reported as:

1. *No growth.*
   Sometimes a urine specimen will produce no growth on culture but will contain pus cells. Causes of sterile pyuria are:
   (a) renal tuberculosis
   (b) antibiotic therapy
   (c) urethritis
   (d) vaginal contamination (usually epithelial cells and Gram-positive bacilli seen)
   (e) presence of fastidious organisms such as mycoplasmas
   (f) L-form of organism
   (g) non-infective causes such as tumours, calculi and catheters
2. *No significant growth* (less than 10^4 organisms per ml, or scanty mixed contaminants).
3. *Many mixed organisms* (indicating faecal contamination of the specimen).
4. *Approximately 10^4 organisms per ml, of doubtful significance.* (If composed of several bacterial species, for example *Staphylococcus epidermidis* (*albus*), lactobacilli or diphtheroid organisms, this probably represents contamination although the detection of 10^4 organisms per ml of a single species may occasionally be significant).
5. *A significant growth* (greater than 10^5 organisms per ml of a single species, or occasionally two species).

In acute infections the growth is usually of a single organism which in some 75% of cases is *Escherichia coli*; in chronic infections the cultures may contain more than one organism and *Klebsiella*, *Proteus* and *Pseudomonas* species and *Streptococcus faecalis* (enterococci) are most common. A few infections are caused by *Staphylococcus epidermidis* (*albus*) and *saprophyticus*.

Prophylaxis of UTI

This can be achieved by long-term administration of antibiotics usually in reduced dose to kill any organisms that enter the bladder from the gut, although the danger is an increase in the resistance of the intestinal flora. Where predisposing causes exist such as renal calculi or obstruction to the outflow of the bladder surgical intervention will be required.

Treatment

Asymptomatic infection needs no treatment except during pregnancy and possibly during childhood. *Symptomatic infection* demands adequate urinary concentration of antibacterial agents to which the organism is sensitive. Useful agents include cotrimoxazole, nitrofurantoin and nalidixic acid. Cephalosporins and ampicillin may be used but there are now many reports of high levels of resistance to ampicillin in *Escherichia coli* particularly in hospital practice. A problem is a high recurrence rate, either by relapse which indicates failure of the initial treatment to eradicate the organism, or reinfection. In patients with indwelling catheters bladder washouts are useful.

Genito-urinary tuberculosis

Tuberculosis of the kidney is a consequence of invasion of the blood by *Mycobacterium tuberculosis*; early lesions occur first in the blood vessels surrounding the glomeruli. These microscopic foci expand, erode into the collecting system and hence into the pelvis, with downward spread to the bladder, and produce tuberculous cystitis. The epididymis and seminal vesicles may also become infected. Frequently routine culture of the urine is negative and in many of these cases a Gram-stained film shows pus cells without organisms. In cases of pyuria without bacteriuria tuberculosis must be considered.

Excretion of tubercle bacilli from the kidney is intermittent and because of this an MSSU is quite unsuitable for laboratory examination. *Instead, at least 100 ml of early morning urine (EMU) should be sent to the laboratory, on 3 consecutive days.* Ziehl-Neelsen films on a spun deposit are examined and culture is performed on Löwenstein-Jensen, modified Löwenstein-Jensen or sodium pyruvate media. Guinea-pig inoculation is a useful procedure in suspected renal tuberculosis.

Treatment

This must be continued for a period of 9 months to 2 years, using modern anti-tuberculous drugs (see Ch. 5).

Glomerulonephritis

Aetiology

Poststreptococcal glomerulonephritis is a common type of acute nephritis but there may be other aetiological agents. In comparison with rheumatic fever which can be caused by a large variety of types of *Streptococcus pyogenes*, acute glomerulonephritis (AGN) is produced by a very much narrower range of serotypes.

Pathogenesis and epidemiology

Human AGN relates both to *infection* and *immunity*; theories as to its production include the following:

Soluble-complex induced. Soluble complexes, or toxic complexes, circulate in the body having escaped the action of the phagocytes. The complexes which consist of antigen, antibody and complement are filtered off in the glomerulus, from which they are generally removed after several weeks. Antigen involved is quite unrelated to the kidney. The hepatitis B surface antigen has been found in the complexes.

Antikidney antibody. (Anti-glomerular-basement-membrane antibody). Normal persons release glomerular-basement-membrane-reacting fragments in both plasma and urine and at times these may become immunogenic.

Pathogenetic mechanisms may also relate to cross reactions between streptococcal membrane antigens and other human membrane antigens, including the glomerular-basement-membrane.

Whichever of the theories is correct—and all may be involved from time to time—the events that follow the combination of antigen and antibody in the kidney are the same, namely *activation of complement* and *coagulation*. The main cause of injury to the glomerulus is caused by activation of complement with the concomitant attraction of leucocytes and immune adherence with release of enzymes from the leucocytes. Complement components, particularly C3, are reduced in AGN. Complement also initiates *coagulation* which leads to local tissue damage through microthrombosis and inflammation. Although the above events are all potentially reversible the healing processes quite often produce scarring.

The role of both skin and throat infections in the development of AGN is now firmly established, although there are clear differences in the epidemiology. For example, AGN following streptococcal throat infection occurs at the colder times of the year and mostly in children, whereas AGN following infected skin lesions (*pyoderma*) such as impetigo occurs in all age groups in the hot humid weather, notably in Alabama and the West Indies, where the Hippelates fly is an important vector of streptococci. The nephritogenic serotypes are also different. AGN following throat infection can be caused by M-types 12, 1, 25, 4 and 3, whereas AGN following pyoderma involves M-types 49 and 52, 53, 54; 55, 57, 58; 59, 60, 61. Recurrences of AGN are rare.

Laboratory diagnosis

Although beta-haemolytic streptococci may be isolated from the throat and skin, serological tests are generally more helpful. Immune responses vary depending on whether the source of infection is the throat or skin. For example, the *antistreptolysin O* (ASO) estimation is reliable in the throat form but unreliable in pyoderma-associated AGN, whereas the *anti-deoxyribonuclease B (anti-DNAse B)* titre is raised more frequently and to a greater degree in pyoderma-associated AGN.

Prophylaxis

By treating skin lesions AGN may be prevented, but treatment of throat infections has little effect.

Treatment

Antibiotics are ineffective in established AGN.

Perinephric abscess

This can arise: (1) by extension of an abscess in the cortex of a kidney; (2) by extension of an appendix abscess via periureteral lymphatics; (3) from the blood. Surgical drainage, with or without antibiotics, is the treatment of choice.

Viral infections

A number of viruses can be isolated from the urine. In most cases, the source is not known, although CMV is known to infect the cells of the collecting tubules. Urine is an excellent specimen for virus isolation from all patients with suspected CMV infection—from the newborn to adults and especially the immunocompromised transplant recipient. It is often useful to remember that mumps virus is excreted in the urine for up to 2 weeks after infection.

BK virus, a human polyoma-like virus causes widespread inapparent infection in more than 50% of the adult population. Virus persists in the host, however, and may be excreted if the immune system is impaired. In transplant recipients the virus has been isolated from the urine and viral antigens have been located in foci of infection around the ureters. Some adenoviruses and *Chlamydia trachomatis* have been associated with cystitis.

FURTHER READING

Asscher A W 1974 Pathogenesis and management of urinary tract infection in women. British Journal of Hospital Medicine 12: 546–554

Asscher A W 1978 Management of frequency and dysuria. British Medical Journal 1: 1531–1533

Cameron J S 1972 Bright's disease today: the pathogenesis and treatment of glomerulonephritis. British Medical Journal 4: 87–90; 160–163; 217–220

Dillon H C 1970 Streptococcal skin infection and acute glomerulonephritis. Postgraduate Medical Journal 46: 641–652

Duerden B I, Moyes A 1976 Comparison of laboratory methods in the diagnosis of urinary tract infection. Journal of Clinical Pathology 29: 286–291

11

Infections of the genital tract

Normal flora of the vagina

The vagina is colonised very soon after birth. In considering the normal flora of the region it is important to remember the various ecological areas in and around the vagina, such as the periurethral glands, Bartholin's glands and the cervical canal. Organisms are not uniformly distributed throughout the female genital tract and some areas support growth of predominantly anaerobic flora whereas others support the growth of the more common aerobes. The biochemical properties of mucus may contribute to the numbers, types and location of bacteria in the female genital tract.

Many different species can be cultured from the vagina depending on the methods used for isolation and on the site sampled. The nature of the flora also relates to the age of the female and to whether intra-uterine devices or tampons are being used. In general there are several groups of organisms commonly present: *Gardnerella vaginalis* (*Haemophilus*) lactobacilli, diphtheroids, beta-haemolytic streptococci, group B streptococci, anaerobic cocci, coliform organisms, *Streptococcus faecalis*, *Staphylococci*, *Bacteroides* spp and yeasts.

Several authors have postulated differences in the vaginal flora of the pregnant and non-pregnant woman but the differences are not constant. Sexual factors as well as physiologocal factors may play a more important role in determining numbers and species of organisms, and because it is not absolutely clear which organisms are pathogens in the vagina it is not strictly possible to describe what is normal flora.

Defences against infection in the female genital tract

The presence of glycogen in the vaginal mucosa, controlled by the secretion of oestrogens and progesterone and acted upon by lactobacilli (Doderlein's bacilli) to produce lactic acid and an environment of pH 5, is an important mechanism of protection in the vagina. This pertains to the newborn, whose hormones are derived from the mother, and to women in the reproductive age group. During childhood and after the menopause when oestrogen activity is absent glycogen is not deposited in the vaginal mucosa and the vaginal pH rises to between 6 and 7.

The flora of the vagina and of the periurethral area prevents colonisation by other species and an intact epithelium is also important in this.

Potential pathogens (see Table 11.1)

Vulvitis

This is commonly caused by *Trichomonas vaginalis*, yeasts and herpes simplex virus but may be caused occasionally by gonococci. *Vulvo-vaginitis* occurs in young girls often as the result of inadequate perineal hygiene but sometimes as a sequel to the presence of some foreign body in the vagina. Staphylococci and coliform organisms are often isolated and occasionally gonococci. *Bartholinitis*, acute infection of the gland and duct of Bartholin in the posterior third of the labia majora, may also be associated with gonococci, coliform organisms and anaerobic bacteria.

Vaginitis

In addition to trichomonal and yeast infections a non-specific form of vaginitis occurs from which staphylococci, streptococci, *Gardnerella*, *Mobiluncus*, anaerobic cocci and other intestinal organisms may be isolated; this may be associated with the use of pessaries, douches and tampons.

Toxic shock syndrome (TSS)

This is caused by certain toxigenic strains of *Staphylococcus aureus* and occurs in a proportion of women who are menstruating and who use highly absorbent tampons. The disease is characterised by pyrexia, confusion, headache, subcutaneous oedema, hypotension and a scarlatiniform rash. Vaginal swabs from such sufferers almost invariably produce growth of *Staphylococcus aureus*.

INFECTIONS OF THE GENITAL TRACT   179

Table 11.1 Potential pathogens of the genital tract

Infection	Organisms
Epididymitis	Coliforms Streptococci (various) Staphylococci *Myco. tuberculosis* *N. gonorrhoeae*
Orchitis	*T. pallidum* Mumps virus
Balanitis	*H. ducreyi* *T. pallidum* *C. albicans* *T. vaginalis* *C. trachomatis* Coliforms
Vaginitis	Beta-haemolytic streptococci (groups A and B) *Staph. aureus* Enterobacteria Candida Trichomonads
Cervicitis	Beta-haemolytic streptococci *Staph. aureus* Bacteroides spp Enterobacteria *N. gonorrhoeae* *T. pallidum* (primary chancre) Herpes simplex virus *C. trachomatis* Papillomavirus Cytomegalovirus
Endometritis (puerperal)	*Strept. pyogenes* Anaerobic cocci *C. perfringens* Bacteroides spp *Staph. aureus* Mycoplasma
Salpingitis	Coliforms Various aerobic and anaerobic cocci Staphylococci *Myco. tuberculosis* *N. gonorrhoeae* *C. trachomatis*
Oophoritis (rare)	Mumps virus Extension of salpingitis

Cervicitis

Acute cervicitis may be *primary* as the result of injury, for example an accidental tear during childbirth, or *secondary* due to inflammatory lesions in the endometrium or vagina. Staphylococci, streptococci, coliform organisms, gonococci, *Bacteroides* spp, as well as trichomonads and yeasts, may be involved, and acute cervicitis may occur in sexually-transmitted diseases such as chancroid and infection with *Chlamydia trachomatis* and herpes simplex viruses.

Endometritis

The myometrium and pelvic peritoneum may also be involved. This occurs most commonly following childbirth and abortions, and as the result of gonorrhoeal infection. A wide variety of organisms may cause infection associated with the use of intrauterine devices (IUD).

Salpingitis

Acute salpingitis may occur: (1) by upward spread of infection from the uterus along the mucosa of the tube, as in gonococcal and *Chlamydia trachomatis* infection; (2) from the pelvic cellular tissue to which infection has spread from the endometrium through the uterine wall, by veins and lymphatics. Here the tube is attacked from the periphery, infection proceeding from outside into the mucosa of the tube, as in post-abortal or puerperal salpingitis; (3) from the bowel as in appendicitis; and (4) by the bloodstream as in tuberculosis.

The ovaries are usually also involved (salpingo-oophoritis), the condition is commonly bilateral and there is frequently an associated pelvic peritonitis. A perihepatitis can occur from the spread of *Chlamydia trachomatis* from infected tubes.

Pelvic cellulitis

This is also known as pelvic inflammatory disease (PID) and is generally combined with pelvic peritonitis. Infection is usually confined to the tissues of the pelvic floor. It may start in one broad ligament and spread to the other, in front of and behind the uterus, or may spread upwards via the cervix; occasionally salpingo-oophoritis may occur. Pelvic cellulitis may follow injury to the upper vagina and cervix as the result of parturition, abortion, pelvic surgery and careless dilation of the cervix; it may follow appendicitis, diverticulitis and infection of a bowel carcinoma. Abscess formation rarely occurs nowadays.

As in all gynaecological infections adequate specimens such as swabs, pus and cervical scrapes should be sent for investigation of bacteria, chlamydiae and viruses. Blood cultures must be taken. When specimens cannot be obtained 'best-guess' antibiotic therapy may be required.

Urethritis

Neisseria gonorrhoeae, although most commonly involved in acute urethritis, can also cause *prostatitis*. *Non-specific urethritis* is produced in over 50% of cases by *Chlamydia trachomatis*. Essential specimens for examination include a urethral swab and pus. *Balanitis* is associated with various organisms and a swab should be taken from inside the prepuce. Herpes simplex virus can also cause urethritis.

Epididymitis

Although infection may be confined to the epididymis, *epididymoorchitis* may also occur.

Infection occurs via the vas and is usually secondary to infection of the urinary tract, prostate and seminal vesicles. It can also occur after prostatectomy and urethral instrumentation.

Gram-negative bacilli, particularly those associated with urinary tract infection, are frequently involved, and to a lesser extent staphylococci and gonococci. *Myobacterium tuberculosis* and mumps virus are infrequent causes.

Culture of urine is essential in diagnosis.

Orchitis

This can occur either as an extension of epididymitis or via the bloodstream as in mumps. Semen has been shown to be infectious in patients with hepatitis B and those carrying HIV. CMV can also be excreted by this route.

SEXUALLY-TRANSMITTED DISEASES

These are the most common communicable diseases after the common cold and their incidence increases yearly. Only a minority of patients suffer from the classical venereal diseases such as syphilis (*Treponema pallidum*), gonorrhoea (*Neisseria gonorrhoeae*), chancroid (*Haemophilus ducreyi*) lymphogranuloma venereum (*Chlamydia*), granuloma inguinale (*Donovania granulomatis*).

The majority of patients suffer from the following:
Non-gonococcal urethritis (NGU) (caused by *Chlamydia trachomatis*).
Trichomoniasis (*Trichomonas vaginalis*)
Genital candidiasis (*Candida albicans*)
Warts (papilloma virus)
Molluscum contagiosum (a pox virus)
Genital herpes simplex
Ectoparasite infestation (crab louse; scabies)

Table 11.2 shows the numbers and types of new cases of sexually transmitted diseases reporting to clinics in the United Kingdom, 1981-83.

Table 11.2 Sexually transmitted disease: reported new cases 1981-83 (Scotland, England and Wales). Information provided by The Communicable Disease Surveillance Centre, Public Health Laboratory Service and the Communicable Diseases (Scotland) Unit.

Diagnosis	1981	1982	1983
Syphilis	4211	3929	3727
Gonorrhoea	58 301	58 778	54 859
Chancroid	100	137	81
Lymphogranuloma venereum	41	38	43
Granuloma inguinale	29	20	23
Non-specific genital infection	132 391	142 072	148 616
Trichomoniasis	21 625	21 517	19 571
Candidiasis	50 954	56 124	62 199
Scabies	2434	2304	2477
Pubic lice	9749	10 904	10 198
Herpes simplex	12 080	14 842	17 908
Warts	33 480	37 341	42 790
Molluscum contagiosum	1305	1467	1700
Other treponemal disease	884	843	746
Other conditions requiring treatment	73 817	85 315	98 230
Other conditions not requiring treatment	121 918	127 208	132 777
Total newcases	523 319	562 839	595 945

Trichomoniasis

This common protozoal infection is caused by *Trichomonas vaginalis* which produces a thick greenish-yellow vaginal discharge. Examination of wet films of pus will reveal the typical motile organisms. Metronidazole is the treatment of choice.

INFECTIONS OF THE GENITAL TRACT

Genital candidiasis (thrush)

This is caused by several species of *Candida* but the most common cause is *Candida albicans* which is often a commensal organism in the vagina. When infection occurs white membranous patches are produced in the vagina and vulva and there is vaginal discharge which may be thick or watery. Gram films identify the yeasts. Nystatin administered topically or in pessaries is used for treatment.

Genital ulcers

Infectious causes include syphilis, chancroid (*Haemophilus ducreyi*) and herpes simplex virus: erosions may occur in association with candida, *Trichomonas vaginalis*, warts, scabies and simple pyogenic infections.

Gonorrhoea

Bacterial aetiology

Neisseria gonorrhoeae (the gonococcus).

Pathogenesis and epidemiology

The gonococcus is a strictly human parasite and unlike the meningococcus is seldom found in carriers, although asymptomatic carriers do occur. The anterior urethra is affected in men and the anterior urethra and cervix in women. Infection is generally limited to the mucosa, although spread in men can affect the prostate, seminal vesicles and epididymis. In women advanced infection may affect the uterus and fallopian tubes although the vaginal mucosa is rarely involved. *Rectal infection* and throat carriage occurs in both sexes. Reports have described a 5-10% carriage rate of gonococci in the *throat*. *Gonococcal ophthalmia*, an acute conjunctivitis in the newborn, may result from infection of the baby during birth and metastatic infections such as arthritis may be produced if the organisms invade the bloodstream. Gonococcal vulvo-vaginitis in young girls is now an uncommon infection except in association with sexual offences.

Laboratory diagnosis

The best specimens from the female are a urethral swab and a cervical swab. High and low vaginal swabs give inferior results. A swab of urethral discharge and of any discharge after prostatic massage is taken from males and rectal swabs and throat swabs may also be obtained. If possible cultures should be taken directly from the patient using a wire loop because the gonococcus is a delicate organism and dies quickly outside the body; alternatively, transport media may be used.

Although the appearance of *intracellular Gram-negative diplococci* from such specimens is virtually diagnostic of gonorrhoea, cultures are essential for medico-legal purposes. Chocolate agar, Thayer-Martin and New York City media incubated in 5–10% carbon dioxide may be used. The *oxidase test* and *sugar fermentation tests* are used for further identification of suspected organisms in pure culture. *Fluorescent tests* for the identification of gonococci in smears may be used but cultural and biochemical confirmation is still required.

The gonococcal complement fixation test (GCFT) is of limited value. False-negative and false-positive reactions are common; the main problem is that of cross reactions with antibody to other commensal neisseriae.

Treatment

The gonococcus is sensitive to a wide range of antibiotics although in recent years a marked increase in resistance to penicillin has occurred. Cotrimoxazole and erythromycin are alternative drugs.

Syphilis

Bacterial aetiology

Treponema pallidum.

Pathogenesis and epidemiology

Almost all cases are contracted by sexual intercourse because like the gonococcus the causative organism is delicate and dies off rapidly outside the body. The spirochaetes present in the superficial genital lesions are passed from one partner to another through intact or damaged mucosae. When the spirochaetes have established themselves at the point of entrance they multiply rapidly and within the next 3 months the *'primary sore'* or *'chancre'* appears, usually on the genitalia. Although the lesion is localised the spirochaetes are distributed widely throughout the body.

The *secondary stage* occurs several weeks after clinical infection and widespread skin rashes and other lesions may occur, such as the characteristic throat and mouth ulcers and condylomata of the anus and vulva. These mucosal lesions discharge large numbers of spirochaetes. About this time the primary chancre disappears leaving only a trace of its presence, as a scar.

The *tertiary stage* is generally delayed for many years and takes the form of chronic granulomata known as *gummata*, in the brain, bone, skin and internal organs. Late manifestations of syphilis are degeneration of the brain cells producing *general paralysis of the insane* (GPI) and destruction of nerve fibres in the spinal cord causing *tabes dorsalis*. Lesions of tertiary syphilis contain few spirochaetes. The massive tissue damage may be accounted for by a delayed hypersensitivity reaction. Infection during pregnancy can be transmitted to the fetus.

Laboratory diagnosis

The organisms cannot be cultivated on artifical culture media. The diagnosis of syphilis is confirmed either by finding *Treponema pallidum* in lesions or by demonstrating antibodies in the serum.

1. Demonstration of Treponema pallidum

Exudate from primary or secondary lesions may be examined by dark-ground microscopy. Although non-pathogenic spirochaetes may be present they are thicker and more irregular in shape, and more motile.

2. Serological tests

   a. Non-specific tests. Antibodies against cardiolipin can be demonstrated by *complement fixation* as in the *Wassermann Reaction*, or by *precipitation (flocculation) tests* such as the *Kahn* and *Venereal Diseases Research Laboratory (VDRL)* tests. The latter test is simple and reliable and is now widely used. It becomes positive early in the infection, usually about 2 weeks after the appearance of the chancre.

These tests are non-specific and false-positive reactions can therefore occur, such as in pregnancy, malaria, glandular fever and in the acute stages of some infectious diseases.

b. Treponema pallidum-*specific tests*. The tests that utilise *Treponema pallidum* (Nichol's strain) as an antigen are (1) the *Treponema pallidum immobilisation test (TPI)*; (2) the *fluorescent treponemal antibody absorbed test (FTA-ABS)*; and (3) the *Treponema pallidum haemagglutination test (TPHA)*.

The *TPI* test, often negative in primary syphilis, is complicated and technically difficult to perform and is used in only a few reference laboratories. The *FTA* test is simple and specific. It is generally used as a reference test when screening tests give divergent results, and is highly specific. The *TPHA* test is suitable for routine screening because it is simple to perform and commercial reagents are available. It is a very useful test although often negative in early untreated primary syphilis.

The VDRL and TPHA tests are the most useful in routine service laboratories and when performed together provide an effective screen for early and late infections.

Treatment

Penicillin in high doses is very effective against *Treponema pallidum*. Erythromycin or tetracyclines may also be used in cases of penicillin hypersensitivity. The Jarisch–Herxheimer reaction may occur when treatment is started.

Chlamydial infections of the genital tract

Aetiology

Chlamydia trachomatis serotypes D–K. Serotype L is the organism associated with lymphogranuloma venereum.

Pathogenesis and epidemiology

Chlamydia trachomatis is a strict intracellular parasite. Infection is acquired by direct transfer during intercourse.

The male and female contract urethritis after an incubation period of a few days: the rectum may also be infected and a few cases of cystitis have been described. In the female, the cervix is an important site of infection. Cervicitis and discharge may develop or often *there may be no sign of infection*. Ascending infection may lead to prostatitis or salpingitis and pelvic inflammatory disease. Gonococcal infection may also be present; pharyngeal infection has also been described.

The organism can be transferred from the cervix to the infant eye, to cause an inclusion conjunctivitis within 1–2 weeks of birth (30–50%) or to the respiratory tract as a pneumonia (10–20%) within 1–3 months of birth. *Lymphogranuloma venereum* is an invasive disease in which infection spreads to the tissues around the urethra, vagina and rectum and to the inguinal lymph nodes, which enlarge and may discharge. Considerable scarring occurs on healing, and strictures and fistulae may develop.

Laboratory diagnosis

Chlamydia trachomatis is inactivated rapidly outside the body and specimens must be collected into special transport medium and either transferred quickly to the laboratory or stored at low temperature. Cultures should be taken from the urethra, cervix, rectum and throat as indicated. Isolation is attempted by spinning the sample onto treated cell cultures and incubating for 3 days. Isolation is confirmed by staining with iodine and microscopical examination for the glycogen-rich inclusion bodies of the sub-group A Chlamydia.

Detection of the elementary bodies in a smear of exudate is now possible using a direct immunofluorescence test with a monoclonal antibody or by ELISA. Superficial infections induce an antibody response but a sensitive immunofluorescence assay is needed to detect them. The invasive infection with *Chlamydia trachomatis* type L is readily diagnosed by a complement fixation test with the group antigen.

Treatment and control

Chlamydia trachomatis is sensitive to the tetracyclines and erythromycin. Relapse or re-infection may occur and 'test of cure' cultures should be taken. It is important to screen and test female contacts as there may be no clinical signs of infection.

Herpes genitalis

Aetiology

Herpes simplex virus, types 1 and 2.

Pathogenesis and epidemidology

Herpes genitalis presents in the male as vesiculation and ulceration of the glans, shaft of penis, rectum and perineal skin, or neighbouring skin of thigh and mons pubis. A urethritis can also occur. In the female, the lesions often develop on the skin of the perineum, thighs and the vulva: in these sites the lesions may fuse and be so painful that micturition is impossible. A radiculitis may be present and contribute to the difficulty in passing urine. The walls of the vagina may be infected; the cervix is an important site, with the production of cervical ulcers or cervicitis.

In primary infections the lesions are more severe and expand over 5 days; pain lasts for about 10 days and healing may take at least 2 weeks; inguinal adenopathy is usually present. In recrudescent disease, the lesions develop over 2–3 days, are painful for about 6 days and heal within 10–12 days; adenitis is unusual. In the genital area the virus is maintained in the sensory root ganglia of the sacral nerves.

Overall most herpes genitalis is caused by type 2 virus. However, a significant number of cases are due to type 1, especially in social groups where many young adults have not acquired the virus in childhood. Thus in female patients up to 30% of genital isolates may be type 1, but the proportion is even higher (50%) in girls under 20 years. In males, up to 20% HSV-1 genital infections have been recorded. Recrudescence is more frequent with type 2 virus and 40% of patients can expect to suffer from this after a primary genital infection. HSV-2 can infect extragenital sites such as the hand, breast and throat.

Complications

Retention of urine in female patients and viral meningitis in a baby born to an infected mother, especially if the mother has a primary infection late in pregnancy. Viral meningitis due to HSV-2 has been recorded in young adults.

Laboratory diagnosis

Specimens submitted should include vesicle fluid, swabs or scrapes in transport medium from lesions and clotted blood for serology studies. Virus may be detected in material from the lesions by electronmicroscopy and immunofluorescence. Virus can be readily isolated in cell culture and a cytopath effect may be visible within 24–48 hours. Serology studies may be useful in primary infections.

Treatment and control

Virus shedding and pain can be reduced by early treatment with acyclovir given either orally for severe primary infections or topically for recrudescent disease. Pregnant women with a history of past or suspected herpetic infection should be screened as delivery approaches. If virus excretion is found the baby may be delivered by caesarean section to avoid transmission, as infection disseminates to the CNS, in which case the mortality rate may be over 50%. Fortunately this complication is rare.

A vaccine for active immunisation is under study.

Carcinoma of the cervix

The epidemiology of this disease indicates that it is a transmissible disease. Important risk factors are the age of first intercourse and pregnancy, and the number of sexual partners. Genital infection with herpes simplex virus, usually HSV-2, has been implicated. However, infection with some human papilloma viruses (HPV) is now recognised as an important factor in the evolution of the disease.

It is now possible to type HPV by hybridisation studies; types 6, 11, 16 and 18 are most often detected in the cervix. The DNA of types 16 and 18 is associated with invasive carcinomas of the cervix, vulva and penis: in such tumours, the viral DNA is integrated in the cell chromosomes. HPV types 6 and 11 are found in benign genital warts or in cervices showing mild dysplasia: in these the viral DNA is episomal.

FURTHER READING

Corey L, Adams H G, Brown Z A, Holmes K K 1983 Genital herpes simplex virus infections; clinical manifestations, course and complications. Annals of Internal Medicine 98, 958–972

Garrod L P, Lambert H P, O'Grady F 1981 Antibiotic and chemotherapy. Churchill Livingstone, Edinburgh

Oriel J D, Ridgway G L 1982 Genital infection by *Chlamydia trachomatis*. Edward Arnold

Robertson D H H, McMillan A, Young H 1986 Clinical practice in sexually-transmitted diseases. Churchill Livingstone, Edinburgh

Schwarz E, Freese U K, Gissmann L, Mayer W, Roggenburk B, Stremlan A, zur Hansen H 1985 Structure and transcription of human papilloma virus sequences in cervical carcinoma cells. Nature 314: 111–114

Singer A, Campion M J, McCance D J 1985 Human papillomavirus. British Journal of Hospital Medicine August: 104–108

Walker J, MacGillvray I, Macnaughton M C 1976 Combined Textbook of Obstetrics and Gynaecology. Churchill Livingstone, Edinburgh

Young H, Henrichsen C, Robertson D H H 1974 *Treponema pallidum* haemagglutination test as a screening procedure for the diagnosis of syphilis. British Journal of Venereal Diseases 50: 341–346

12

Infections in pregnancy and the puerperium: congenital and neonatal infections

Infections in pregnancy

Any such infection may cause spontaneous abortion, premature labour or stillbirth. Risks to the fetus greatly outweigh those to the mother because of its immature phagocytic and immune defence mechanisms.

In early pregnancy virus infections of the mother such as rubella, mumps, measles and cytomegalovirus infections may produce spontaneous abortion. Bacterial infections are not common at this stage and, with the exception of those caused by *Treponema pallidum* and *Listeria monocytogenes*, do not appear to cause fetal damage. Septic abortion is now rare in Western Countries because of abortion law reform.

In later pregnancy, intra-uterine infection such as chorioamnionitis can be produced by organisms that colonise the mother's genital tract and perineum, such as group B streptococci, *Listeria*, coliforms such as *Klebsiella*, *Pseudomonas* and *Escherichia coli*, and anaerobes. *Mycoplasma hominis* and herpes simplex virus may also initiate infection via this ascending route.

INFECTIONS IN THE PUERPERIUM

Puerperal pyrexia

Puerperal pyrexia is defined in Scotland as a temperature of 38°C (100°F) sustained for 24 hours, or recurring within that period, during the 21 days after childbirth. In England and Wales 14 days are postulated. Sources of pyrexia are most commonly genital, breast, urinary or respiratory infection.

Puerperal sepsis (Childbed fever)

Bacterial aetiology

Many bacteria can produce infection in the postpartum uterus, most commonly *Staphylococcus aureus*, beta-haemolytic streptococci, anaerobic cocci, *Streptococcus faecalis*, *Bacteroides* spp and coliform organisms.

Pathogenesis and epidemiology

Infection can be endogenous or exogenous. Endogenous sources are the upper respiratory tract, skin and hair of the patient where there may be staphylococci and streptococci, and the alimentary tract via the perineum from where infection may be caused by the anaerobic cocci and enterobacteria. Exogenous sources are either similar sites in personnel looking after the patient, or environmental, and may be airborne, transferred by infected hands, or the result of careless instrumentation. Puerperal sepsis is basically a wound infection with consequent inflammatory reaction in the pelvic organs. Nowadays infection rarely spreads from the placental site to the tubes and ovaries because of antibiotic therapy.

Laboratory diagnosis

High vaginal swabs and particularly cervical swabs may be helpful but pus or discharge should be processed if obtainable. Care should be taken to ensure that sufficient media are used to facilitate isolation of the more fastidious organisms, particularly anaerobes. Blood must be taken for culture.

Treatment

This will depend on laboratory reports although it is often necessary to start treatment on an empirical basis until accurate information can be obtained. Combined therapy with a penicillin or aminoglycoside with metronidazole (as infection is often anaerobic) is generally effective.

Breast abscess

This can occur in the puerperium or may be delayed for days or weeks. It is caused by *Staphylococcus aureus* entering the breast via the nipple.

Congenital infections

Contracted by the transplacental route these result in damage to various fetal organs, or in abortion or stillbirth. The principal pathogens are *Treponema pallidum*, *Toxoplasma gondii*, rubella and cytomegalovirus. Intrauterine infection with hepatitis B virus may occur but is more likely to be acquired at birth or soon after. HIV may be acquired in utero.

Congenital syphilis

Routine antenatal screening and penicillin treatment of positive cases has made this a very rare disease in the West. Many body systems can be involved such as the central nervous system, skin, bone, cartilage and liver.

Congenital toxoplasmosis

Toxoplasma gondii causes choroidoretinitis and encephalitis with sequelae such as mental retardation and impaired vision.

Congenital rubella

Pathogenesis

Rubella, or german measles, is a mild disease in children and adults. Virus is inhaled, infects and penetrates for further growth in cells of the spleen and lymph nodes before spreading via the blood to the skin. The incubation period is from 2–3 weeks; virus is present in the throat for some days before and for about 1 week after the onset of symptoms of fever, rash, cervical and occipital adenopathy. In adults there may be a transient arthritis. The importance of rubella arises from its ability to infect the placenta and hence the fetus. If this occurs in the first weeks of pregnancy fetal death may result or malformations of heart, eye and ear may be present at birth. Further features include mental retardation, hepatosplenomegaly, and low birth weight. Minor damage to the ear may not be detected for some years. The risk to the fetus is greatest in the first 2 months. Thereafter the risk diminishes progressively and is no more than 1–2% by the fifth month. Rare complications are diabetes and panencephalitis which develop after birth. It is essential to make an accurate diagnosis of rubella occurring in pregnancy in order that the risk to the fetus can be assessed. Virus isolation is difficult and slow, therefore the diagnosis is made by demonstrating a significant rise in antibody titre, or the presence of IgM antibody to rubella virus.

Diagnosis

Most affected infants excrete virus from the respiratory tract and this may persist for up to 1 year. Affected babies have IgM in the blood and this declines over the first year of life to be replaced by IgG antibodies. If infection is first suspected at birth serum samples from both mother and baby should be examined.

Immunisation

Live attenuated vaccines have been developed and are now widely used. They are given by subcutaneous injection; some cause fever and arthritis. In the United Kingdom vaccine is given to all girls aged 11–14 years. As this policy does not aim to eradicate the wild virus unvaccinated women are at risk and infection in pregnancy still occurs.

Although there is no evidence of teratogenicity, vaccine is not given in pregnancy and if given to older non-immune women they should avoid pregnancy for 2–3 months.

Congenital cytomegalovirus (CMV) infection

Pathogenesis

CMV infection is widespread in children and usually silent. A persistent infection results and reactivation is frequent in the immunosuppressed. The virus is excreted in semen, saliva and urine. During pregnancy CMV can be isolated from 3–5% of patients, from the cervix. Infection of the placenta and fetus is most likely to occur during a primary infection which occurs in about 1% of pregnancies: 0.2% of newborns are infected; infection can occur at any stage of pregnancy. The most serious effects are hepatitis, purpura, anaemia and deafness. CNS involvement with ventricular calcification and microcephaly leads to mental retardation. It is estimated that 200–300 affected children are born each year in the United Kingdom. Many congenital infections will not be recognised until some years later.

As with rubella, affected children excrete virus and produce antibody.

Diagnosis

Virus isolation is often slow (from 2 to 4 weeks) but should be attempted from specimens of urine and respiratory secretions. Primary infection can be diagnosed by a rise in antibody titre and the presence of virus-specific IgM. Both mother and baby should be tested.

Control

There are no effective antiviral agents and no vaccine for active immunisation is available.

Perinatal infections

These can occur either in late pregnancy or during labour and delivery and are mostly caused by the same organisms involved in intra-uterine infection, in particular group B streptococci and coliforms, via the ascending route (see above). In addition, sexually transmissible organisms can be of importance, notably *Neisseria gonorrhoeae*, *Chlamydia trachomatis* and herpes simplex type II.

Neonatal bacterial infections

Organisms associated with perinatal infections can also cause infection in the neonatal period, some more commonly than others. These are staphylococci, Gram-negative bacilli, notably *Escherichia coli*, *Pseudomonas aeruginosa* and *Klebsiella aerogenes*, and group B streptococci.

Staphylococcal infections

30–40% of babies become colonised by *Staphylococcus aureus* in the first week of life, particularly in the umbilical area, groin, axillae and nose, and a number become infected. Skin infections include pustules and abscesses. 'Sticky eye' may be caused and the umbilical stump may be infected. More serious staphylococcal infections are rare.

Group B streptococcal (GBS) infections

Lancefield's group B streptococci (*Streptococcus agalactiae*) can be very dangerous pathogens in the neonatal period. In the neonate, GBS infections are termed 'early' or 'late' but these terms are not uniformly defined. Whereas the 'early-onset' term for septicaemic infection is correct, because this occurs within several hours of birth, the timing of 'late-onset' infection has proved highly arbitrary. It is generally agreed, however, that GBS infection occurring beyond the immediate neonatal period falls into the 'late-onset' category. With increasing age and maturity of the infant, 'late-onset' infections which may be days or weeks after birth are expressed more as localised infection of target sites, such the meninges, bones and joints.

'Early-onset' infection is an acute undifferentiated, fulminating sepsis during the first 2 days, and generally within 24 hours of birth. As with meningococcaemia, it causes severe shock. The lung is the site of initial and major involvement and in those dying within a few hours of birth the inflammatory response is minimal or absent. In the majority of cases the neonate appears to become infected by aspirating infected vaginal secretions, but GBS can occur in babies delivered by caesarian section, implying that GBS can enter the amniotic fluid through intact membranes. As with many other neonatal infections predisposing factors to this form of GBS infection are low birth weight, prematurity, prolonged rupture of the membranes, a maternal vagina colonised by GBS and the degree of vaginal colonisation.

Purulent meningitis is the prototype of 'late-onset' infection and this can occur in otherwise healthy full-term infants, born to mothers who are not GBS carriers. The organisms are thought to be acquired from other persons, in contact with the baby, who are GBS carriers.

Coliform infections

Gastroenteritis in babies is serious because of its potential rapid spread in nurseries and because of the water and electrolyte depletion produced. Enteropathogenic (toxigenic) strains of *Escherichia coli*, salmonellas and campylobacters are frequently implicated.

Chlamydia trachomatis

The cervix can infect the eye causing 'sticky eye' within 1–2 weeks of birth. The organism is usually present in the pharynx at the same time and a pneumonia can result.

Neonatal viral infections

Hepatitis B

Occurs usually at birth but may be intra-uterine.

Herpes simplex virus

Infection occurs from the mother's cervix, from oral lesions of the mother or attendants.

Gastroenteritis virus

Rotaviruses may infect in the first month but may not produce symptoms.

Respiratory viruses

RS virus and parainfluenza are important.

Enteroviruses

These viruses are acquired from the mother or attendant. Serious outbreaks have occurred with fatalities. They cause systemic infections, for example Coxsackie and echovirus myocarditis, hepatitis and meningoencephalitis. Immunoglobulin may be of some value in controlling such outbreaks.

Varicella zoster

Maternal chickenpox late in pregnancy may be serious for the neonate if born less than 5 days after the onset, when maternal antibody may not have been produced and crossed the placenta. Hyperimmune zoster immunoglobulin should be given to the baby. Zoster late in pregnancy should not put the child at risk as the mother's antibody levels rise rapidly. However, the mother with zoster can be a source of infection to her attendants and to other babies.

Neonatal fungal infections

Candida albicans

This yeast may cause the infection 'thrush' which usually occurs in the mouth.

PREVENTION OF NEONATAL INFECTIONS

Such infections will never be totally prevented but certain measures help to reduce the incidence. Frequent and thorough hand washing by staff, with the use of antiseptic preparations for the skin (for example, chlorhexidine) and proper attention to aseptic techniques are essential, as is the prevention of overcrowding in nurseries.

Hexachlorophane powder can be used for the babies' skin. In addition specific measures may have to be taken for certain viral infections (see above).

Breast feeding provides the neonate with some protection against gastrointestinal infections in particular, because breast milk contains IgA which with lactoferrin is active against *Escherichia coli*. It also has a high lactose content which encourages growth of lactobacilli to the detriment of opportunistic pathogens.

FURTHER READING

See chapter 11

Hurley R 1983 Virus infections in pregnancy and the puerperium. In: Waterson A P (ed) Recent advances in clinical virology 3. Churchill Livingstone, Edinburgh.

13

Infections of the central nervous system

CENTRAL NERVOUS SYSTEM (CNS)

Defences against infection

Infecting agents reach the CNS from the blood, by direct invasion or by ascending infection of nerves. There are no local defences and hence infection rapidly becomes generalised when an organism gains access to the meninges, subarachnoid space and the cerebrospinal fluid.

Acute meningitis

Meningitis is caused by a wide range of bacteria and viruses (Fig. 13.1). A clear indication of the aetiology of meningitis can be obtained by examination of the cerebrospinal fluid (CSF). (see Table 13.1)

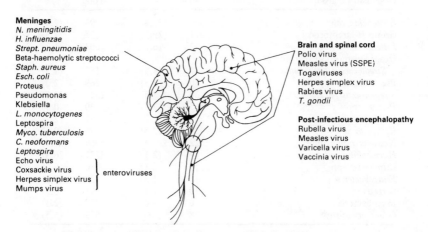

Fig. 13.1 Potential pathogens of the central nervous system

Table 13.1 The cerebrospinal fluid in meningitis

	Normal	Acute pyogenic	Tuberculous	Aseptic
Appearance	Clear	Turbid	Clear or opalescent	Usually clear
Total protein	0.15–0.4 g/l	Greatly increased; (10-fold and more)	Increased	Slightly increased
Sugar	2.6–4.3 mmol/l	Greatly reduced or absent	Reduced	Normal
Cell count	0–3 lymphocytes per mm^3	Greatly increased; $_a$NPNLS	Increased; mainly lymphocytes but some NPNLS	Increased; lymphocytes

$_a$ Neutrophil polymorphonuclear leucocytes

Acute pyogenic meningitis

Bacterial aetiology

Table 13.2 shows the causes of acute bacterial meningitis in England, Wales and Ireland between 1974 and 1983. The aetiology

Table 13.2 Causes of bacterial meningitis in England and Wales, 1975–1983. This information was published by the Communicable Disease Surveillance Centre, London, from data provided by statutory notifications and by voluntary laboratory reports to the Surveillance Centre.

Organism or group	Neonates Total	(%)	All other ages Total	(%)
Escherichia coli	307	(33.4)	293	(2.4)
Group B streptococci	271	(29.5)	157	(1.3)
Listeria monocytogenes	62	(6.7)	203	(1.7)
Other streptococci (except *Streptococcus pneumoniae*	32	(3.5)	157	(1.3)
Pseudomonas spp	30	(3.3)	269	(2.2)
Streptococcus pneumoniae	30	(3.3)	2512	(20.5)
Proteus spp	25	(2.7)	60	(0.5)
Staphylococcus aureus	25	(2.7)	331	(2.7)
Klebsiella spp	23	(2.5)	95	(0.8)
Neisseria meningitidis	22	(2.4)	3969	(32.4)
Haemophilus influenzae	19	(2.1)	3221	(26.3)
Citrobacter spp	19	(2.1)	10	(0.08)
Enterobacter spp	17	(1.8)	20	(0.2)
Serratia	10	(1.1)	15	(0.1)
Mycobacteria	0	(—)	277	(2.3)
Other organisms[a]	28	(3)	679	(5.5)

[a] Includes: coagulase negative staphylococci (9), *Salmonella* spp (8), coliforms (6), *Acinetobacter* (3), *Flavobacterium* (1), *Clostridium* (1).

of acute pyogenic (acute bacterial) meningitis is related to the age of the patient; coliform bacilli and group B streptococci (*Streptococcus agalactiae*) are common causes in the neonate, *Haemophilus influenzae* and *Neisseria meningitidis* in children up to 5 years of age, *Neisseria meningitidis* in older children, adolescents and adults, and *Streptococcus pneumoniae* in adults. *Haemophilus influenzae* may occasionally cause meningitis in adults, and other rare causes include *Streptococcus pyogenes*, staphylococci, salmonellae and *Listeria monocytogenes*.

Pathogenesis and epidemiology

Neonatal meningitis. Although in former years this was commonly caused by *Streptococcus pyogenes* and *Staphylococcus aureus*, coliform bacilli have been an important cause in the last two decades. In recent years, group B streptococci have been responsible for many cases of neonatal meningitis and at present they rival the coliform organisms in being the most common causative agents. Coliform meningitis often results from congenital deformities such as spina bifida, and the sources of the organisms can be the genitourinary tract, lungs or umbilicus. Two forms of group B streptococcal meningitis are recognised (see Ch. 12).

Haemophilus meningitis. Although adult cases have been described this occurs almost exclusively in young children between the ages of 3 months and 5 years. Protection before 3 months is provided by maternal antibody and after 5 years by acquired immunity. Infection is spread by the bloodstream from some focus such as an upper respiratory tract infection. Capsulate organisms of Pittman's type b are most commonly responsible for these infections.

Meningococcal meningitis. Meningococci enter the body via the nasopharynx where they may either produce a localised inflammatory reaction or remain quiescent. There are two main theories as to how they reach the meninges: (1) that they spread by direct extension along the spaces between the sheath and the branches of the olfactory nerve that pierce the cribriform plate; and (2) the more likely theory, that the organisms invade the bloodstream and either produce a transient bacteriaemia or multiply in the blood and cause metastatic lesions elsewhere in the body, in the skin, adrenal glands, joints and meninges. Nine groups of meningococci (A-D; X, Y, Z, Z^1, W135) are recognised; group B is the most common in the United Kingdom.

Infection is most readily spread in conditions of close household contact, although a striking feature of any epidemic is the low morbidity; fewer than 1% of people at risk are infected. An epidemic may be heralded by a rise in the normal 5–10% carrier rate in the community but an increase in the carrier rate does not necessarily mean that clinical infection will occur.

Pneumococcal meningitis. This may be caused either by septicaemic spread secondary to conditions such as lobar pneumonia or otitis media, or by direct infection from the paranasal sinuses following injury to the skull. Infection after injury is a particularly serious form of meningitis in the adult and carries a high mortality rate in elderly patients.

Meningitis due to other bacteria. These include *Staphylococcus aureus* and *Streptococcus pyogenes*. They reach the meninges either via the bloodstream, by direct extension from the ear or paranasal sinuses, or directly from the exterior after trauma.

Tuberculous meningitis. This occurs in a particularly severe form in young children and is probably secondary to a tuberculous focus elsewhere in the body, although this often cannot be located. The bacilli reach the meninges via the bloodstream.

Laboratory diagnosis of meningitis

Laboratory diagnosis is made by the isolation and identification of the specific organism from the blood and cerebrospinal fluid and in some cases by the detection of antigen in the CSF by techniques such as counterimmunoelectrophoresis. It is essential to check for signs of increased intracranial pressure before removing CSF by lumbar puncture.

Blood cultures are particularly useful in the diagnosis of meningitis and may be positive in approximately 50% of cases. About 5 ml of cerebrospinal fluid is removed by lumbar puncture and a cell count made. The fluid is centrifuged and Gram films are made from the deposit (Ziehl-Neelsen films or fluorescent microscopy, if tuberculous meningitis is suspected). Microscopy may be sufficient to indicate the diagnosis and to allow chemotherapy to commence but cultures are also made on blood agar, 'chocolate' (heated blood) agar and on New York City medium which is selective for neisseriae. They are incubated up to 24 hours in an atmosphere of 10% carbon dioxide. Biochemical and/or serological tests identify the organism. For tuberculous meningitis Löwenstein-Jensen slopes are set up and guinea pigs may be inoculated.

Prophylaxis

This applies particularly to meningococcal meningitis. Chemoprophylaxis has been attempted for many years but without a great deal of success. Formerly sulphonamides were used to remove the organisms from the nasopharynx and were effective until a large proportion of meningococci became resistant. The penicillins which are effective and indeed essential in treating the clinical case are ineffective in eliminating the carrier state. Other antimicrobial drugs such as tetracyclines, including minocycline, and rifampicin have been used with varying degrees of success.

Resistance to meningococcal meningitis is related to the possession of antibody. Immunisation with a purified carbohydrate vaccine is possible.

Treatment

Chemotherapy will depend on the nature of the organism and laboratory sensitivity tests, but in general benzyl penicillin is a useful agent; chloramphenicol is used for the treatment of *Haemophilus* meningitis. Sulphonamides are now less widely used empirically because of the problem of increasing bacterial resistance, particularly in meningococci. In general terms a single antimicrobial agent is advocated except in the neonatal period and in the treatment of tuberculous meningitis.

Viral meningitis

Viruses are the most frequent cause of meningitis. The changes in the CSF are indicated in table 13.1.

Aetiology

Most cases are due to enteroviruses—echo- and coxsackie viruses. Many other viruses are associated including HSV-2, togaviruses and lymphocytic choriomeningitis virus. Mumps virus can cause meningitis in children.

Enteroviruses. The human enteroviruses were originally classified as the echo, Coxsackie A and B and polioviruses. New virus isolates are now allocated the next number, for example enterovirus type 71. Now that poliovaccines are widely used, CNS disease due to the three poliovirus serotypes is rare, although occasional cases do occur, usually contracted by the unprotected while abroad. All three polioviruses can cause meningitis, but the commonest agents are echoviruses; types 4, 6, 9, 11 and 30 have been prevalent over recent years in annual epidemics of one type. Coxsackie B viruses and some Coxsackie A (A7, A9) viruses are other common causes, although most A viruses are difficult to isolate and are therefore under diagnosed.

Pathogenesis

Virus enters by the mouth and establishes a silent initial focus somewhere in the nasopharynx or small bowel. Virus then spreads via the lymph and lymph nodes to the blood. Further replication in cells of the lymphoreticular system leads to the major viraemia detectable about 5 days after infection. The virus is free in the plasma. From the blood, virus infects the meninges and may spread via the CSF. At this time fever and neck stiffness appear. Soon after, virus is present in the throat and gut. The illness is benign and symptoms wane within a few days, recovery being complete. The serum antibody titre increases about this time.

Diagnosis

Specimens of CSF, faeces and throat secretions should be submitted to the laboratory for virus isolation. This is readily achieved during the acute phase, although in convalescence faeces are the best source of the virus which may be excreted for some weeks. Enteroviruses grow well in culture and an isolate should be evident within a week. Serological diagnosis is difficult due to the large number of different serotypes of enterovirus. Mumps virus can be isolated from the CSF, throat and urine.

Control

There is no specific therapy. A togavirus, louping ill virus, is endemic in parts of the United Kingdom. It was first recognised in sheep, although the natural cycle is in wild life with a tick as vector. Human cases have been recorded in shepherds and workers preparing vaccine for veterinary use.

Paralytic disease

Aetiology

Polioviruses 1, 2 and 3 and some other enteroviruses.

Pathogenesis

This is similar to that described for echovirus meningitis, except that the illness is typically biphasic. The patient may present with fever and malaise at the time of viraemia; there may then be a brief recovery until the virus invades and destroys cells in the anterior horn of the cord. Paralysis of the muscles ensues. The virus is lytic and the function of the infected cells is not recovered; meningitis may also be present. It must be emphasised that more than 80% of patients infected with wild-type poliovirus suffer no more than a mild feverish illness. As with the other enteroviruses, virus is excreted in the stool for some time after infection. In young children, paralytic disease is rare (infantile paralysis). However, if the age of infection is delayed, the incidence of paralytic disease is increased. Infection during pregnancy carries a high risk of bulbar paralysis. Also, it is known that previous tonsillectomy, injection of alum-absorbed vaccines and muscular activity all increase the risk of paralytic disease. Blood-borne invasion of the CNS is the norm, but entry via nerves may explain some of these features.

Diagnosis

As for enterovirus meningitis, except that virus cannot be isolated from the CSF.

Control

Vaccines both killed and live, are available and very effective. The live vaccine spreads from the vaccinee and with successive cycles of infection there is a risk of reversion to neurovirulence. This may occur about once in 1–2 million vaccines and is usually linked with the type 3 component of the vaccine. Patients who lack an effective humoral antibody response are also at risk of CNS invasion by the live vaccine.

Encephalitis

This may occur alone or accompanied by meningitis and myelitis.

Aetiology

Herpes simplex virus, some enteroviruses, togavirus and rabies virus.

Pathogenesis

Clinically encephalitis presents either as a slowly progressive alteration in behaviour leading to coma, or with the abrupt onset of focal neurological signs. In all cases the prognosis is poor and even if the patient survives there may be considerable neurological damage. Herpes simplex virus, type 1, is the most commonly identified cause in the United Kingdom and there may be 40–60 cases per annum. There is a wide age distribution but most patients are over 25 years of age. Some cases show rising antibody titres suggesting a primary infection but many do not, indicating that these cases have resulted from reactivation of virus either in the brain itself or in sensory ganglia. The route to the brain in such cases is probably by ascent of the cranial nerves. The temporal lobe is usually affected, and this area may contain virus even in cases with diffuse features.

Diagnosis

This is extremely difficult to confirm in the early stages. Brain tissue is the best sample for electronmicroscopy, immunofluorescence and culture as the virus is seldom isolated from lumbar CSF. Serological studies of simultaneous serum and CSF samples may show that there is antibody synthesis within the brain.

Treatment

Untreated there is a mortality of 75%, but this can be reduced by the prompt use of antiviral therapy. The most successful drugs are adenine arabinoside and the present choice, acyclovir. Due to the difficulty in confirming the diagnosis in the laboratory, intravenous administration of acyclovir should be started on clinical grounds although specimens for virus identification and serological tests should be collected.

Togaviruses are important agents in many parts of the world, but not in the United Kingdom. Many alpha- and flavi- viruses can infect man via the bite of an infected mosquito or tick. Man is not a natural host, the virus being maintained in an animal or bird reservoir. The encephalitides are important in the Americas, Europe, Australia and the East, for example Eastern and Western equine encephalitis viruses in the United States. Japanese B encephalitis and Murray Valley encephalitis are also important. As with many other viruses, mild infections are frequent, although the proportion varies with the different viruses. Due to the dependence on the insect vector, epidemics may be seasonal, for example in summer and autumn in the United States.

Diagnosis

Isolation of viruses is difficult and may be dangerous; suckling mice are inoculated. Specimens of CSF and throat swabs should be examined and blood submitted for serological tests.

Post-infectious encephalomyelitis

This serious complication is rare and occurs shortly after infection with measles, rubella and varicella zoster viruses. Formerly it was the most serious complication of smallpox vaccination. In this condition virus cannot be isolated from brain tissue and the disease is believed to result from immunological injury triggered by the virus.

Rabies

Aetiology

Rabies virus, a rhabdovirus.

Pathogenesis

Rabies virus is excreted in the saliva of infected animals. Man is infected by bites, although not all bites from rabid animals will result in infection. Perhaps only about one in five will do so but the risk depends on the extent of the injury; the risk is increased if the bite is on the head or neck. The virus may infect muscle tissue at the site of the bite, but then enters small nerves and ascends to the cord and to the brain. The incubation period is long after bites on the limbs, up to 3 months, even 1 year, but shorter after wounds on the head and neck, reflecting the distance the virus has to travel to reach the brain. Clinically the patient presents with excitement, muscle spasms and hydrophobia due to painful contraction of oesophageal muscles on swallowing, convulsions and coma. Salivary contamination of skin and mucous membranes may be sufficient to introduce virus. Only a few cases of infection by inhalation have been recorded.

Man is not the natural host for the virus which is maintained in wildlife by carnivores such as foxes, jackals and wolves. Skunks and raccoons may be hosts and indeed the virus can infect any warm-blooded animal. This infection is fatal except in South American vampire bats which survive and excrete virus.

Man is often infected by dog-bites although cats can be a source of infection too, especially when their behaviour is altered by the virus. Wild animals may also attack while in the furious stage of rabies, whereas in the dumb phase they may allow themselves to be approached by man. Farm animals such as cows can be infected.

The disease is endemic in most of the world, exceptions including the United Kingdom, Australia and New Zealand. In Europe the disease has been spreading westwards and northwards for some years towards the English Channel.

Diagnosis

Once clinical features develop in the victim viral antigens may be detected by immunofluorescent examination of biopsy tissue rich in nerve endings; suitable specimens are biopsies of skin and corneal smears.

Treatment

In assessing the need for treatment several factors must be considered; for example the time since exposure, the species of animal, its behaviour and the nature of the wound inflicted. It may be possible to examine the animal and if it survives for 10 days there should be no danger. However a more rapid diagnosis can be made by immunofluorescence examination of its brain tissue.

(1) *Post exposure*. In all cases the wound should be cleaned with soap and water and 40–70% alcohol and iodine (tincture or solution) should be applied. The virus is sensitive to these agents and this is an important step, provided it is done promptly. If there is a proven risk initial treatment is followed by (a) passive immunisation, half of the dose given by intramuscular injection, half infiltrated around the wound; and (b) active immunisation with cell culture vaccine.

(2) If clinical symptoms develop intensive care will be required but the outcome is always fatal.

(3) Active immunisation for those at risk, for example those who work in quarantine kennels.

Control

In the United Kingdom strict enforcement of quarantine laws (6 months quarantine) is essential to prevent the entry of the virus and its spread to wildlife.
If the virus became established strict control of dogs would be necessary.

Herpesvirus simiae (B virus)

This is a rare disease of man acquired by the bite of an infected monkey, or by accidental injury while working with cell cultures derived from such animals. The disease is benign in the natural host but in man produces an ascending infection via peripheral nerves to cord and brain. It is almost always fatal.

Cerebral abscess

Bacterial aetiology

The following bacteria can cause this condition: *Streptococcus milleri*, beta-haemolytic streptococci, *Streptococcus pneumoniae*, *Staphylococcus aureus*, *Haemophilus influenzae*, coliform bacilli, *Bacteroides* spp, *Fusobacterium* spp, *Veillonella* and other anaerobic cocci. Infections are often mixed.

Pathogenesis and epidemiology

Abscess formation can occur via the bloodstream or as the result of direct extension into the brain. Abscesses may be extradural, subdural or intracerebral. The sources of infection vary and include otitis media, sinusitis and metastatic spread from chest infection. Infection may also follow dental, abdominal or neurological surgery and may occur after head injuries. The chest is also an important distant site of infection accounting for around 20% of cases. About 10% of abscesses are associated with congenital heart disease. Temporal lobe or cerebellar hemisphere abscesses are associated with otitis media and mastoiditis and frontal lobe abscesses with infection of the ethmoidal, sphenoidal and frontal sinuses. Frontal lobe abscesses are also a complication of acute and chronic dental infection. The mortality rate from brain abscess is about 40% despite treatment and some degree of permanent disability will occur in another 40%. Cerebral toxoplasmosis is recognised as a complication of AIDS.

Laboratory diagnosis

It is most important to obtain pus and make a Gram film in addition to culturing the specimen both aerobically and anaerobically.

Treatment

Where practicable excision of abscesses is mandatory and antibiotics must be given. The penicillin group and gentamicin are widely used, and metronidazole should always be considered because of the frequent associations of *Bacteroides fragilis* and other anaerobic bacteria with cerebral abscess.

Botulism

Bacterial aetiology

Botulism is a severe and frequently fatal form of poisoning caused by the consumption of food containing preformed toxin of *Clostridium botulinum*. Six main types of organisms (A–F) are recognised on the basis of their antigenically distinct toxins.

Pathogenesis and epidemiology

Clostridium botulinum, an anaerobic spore-forming Gram-positive bacillus, is a widely distributed saprophyte in soil, fruits, vegetables leaves, animal manure and sea mud. Human cases of botulism are rare and have been traced to a variety of foods such as fish, ham, sausage, home-canned meats and home-bottled vegetables such as peas and beans. Man is mainly affected by type A but a few cases of type E have been traced to preserved fish. The toxin affects the cholinergic system and seems to block release of acetylcholine. Neurotoxic effects include oculomotor and pharyngeal paralysis, dysphagia and respiratory failure.

Laboratory diagnosis

Suspected food should be examined bacteriologically and Gram-stained films can be examined for sporing Gram-positive bacilli; *Clostridium botulinum* can be isolated using anaerobic techniques; food can be tested for toxin and the patient's blood screened by animal inoculation and toxin neutralisation tests.

Prophylaxis

Home preservation of foodstuffs such as meat, fish, and vegetables such as peas and beans should be avoided, although acidic fruits can be bottled safely since *Clostridium botulinum* does not grow at a low pH. A prophylactic dose of polyvalent antitoxin may be given to asymptomatic persons who are thought to have ingested the toxin.

Treatment

Polyvalent antitoxin is of little value once symptoms have appeared and general supportive treatment is all that can be offered to such patients.

Tetanus

Bacterial aetiology

Clostridium tetani.

Pathogenesis and epidemiology

Tetanus is the result of contamination of a wound with *Clostridium tetani* spores, frequently from manured soil, contaminated dust or clothing. Germination of the spores in a wound depends on reduced oxygen tension which can occur in any devitalised tissue and which may also be produced if aerobes multiply concomitantly in the wound. Infection remains localised although the clinical condition is caused by the effects of a potent diffusible toxin on the nervous system. The toxin travels by way of the motor nerves to the brain and spinal cord and interferes with the transmission of motor nerve impulses.

Although tetanus is commonly associated with deep wounds it can nevertheless result from superficial abrasions, including scratches and rose-thorn pricks.

Laboratory diagnosis

Gram films can be made from the wound exudate and examined for *drum-stick bacilli*. Special stains can reveal the characteristic terminal spore. The bacilli grow well on blood-containing media under strict anaerobic conditions. Biochemical tests can confirm their identity and animal inoculation can test for toxigenicity of the organism.

Prophylaxis

1. *Prevention of germination of tetanus spores.* This includes prompt and thorough wound toilet and surgical debridement if required.
2. *Provision of sufficient antitoxin to neutralise any tetanus toxin that may be produced.* This can be done either by giving (a) a booster dose of tetanus toxoid to those already actively immunised or (b) human tetanus immunoglobulin (HTIG) or (c) anti-tetanus serum (ATS) to those not actively immunised. HTIG should be used in preference to ATS because of the risk of hypersensitivity reactions in patients given horse antiserum. HTIG is freely available in the United Kingdom but this situation may not apply in developing countries, and ATS may have to be given. Combined active-passive immunisation must be given to the non-immune person; this gives immediate passive protection by the HTIG while active immunity is developing.
3. *Prevention of multiplication (and therefore toxin production) of the vegetative bacilli.* This can be achieved by giving prophylactic antibiotics (and can replace 2b), although the use of both procedures in recommended. Antibiotics are no substitute for surgical toilet.

A suggested scheme for prophylaxis is outlined in Fig. 13.2.

Treatment

As well as general measures such as the use of muscle relaxants, sedation and artificial ventilation, antitoxin can be given although this is of little use if the toxin has become fixed to the tissues. Those who recover from tetanus require a course of active immunisation because the disease does not produce immunity.

Chronic virus infections of the CNS

Aetiology

(a) Conventional viruses—measles, rubella, papovaviruses and HIV.
(b) Unconventional agents—Kuru and Creutzfeld-Jakob disease.

Pathogenesis

These are all diseases with long incubation periods (years) and all

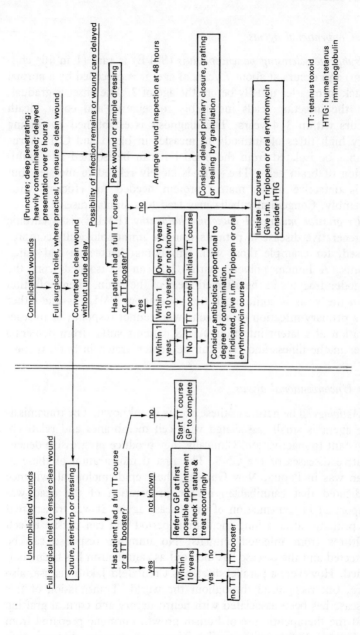

Fig. 13.2 Wound management including tetanus prophylaxis schedule

progress slowly. They are examples of 'slow' or 'slowly progessive' virus infections.

(a) *Conventional agents*

Subacute sclerosing panencephalitis (SSPE) is rare (1 in 10^6 children) and occurs at about 7 years of age; it is preceded by a normal attack of measles usually before the age of 2. The onset is gradual, but the infection leads inexorably to seizures and coma; death occurs within 3–4 years. The diagnosis is established by finding very high titres of antibody to measles in blood and CSF. Virus cannot be isolated from the brain but can be isolated after cultivation of brain cells. The virus is closely related to measles, but it is defective in its matrix protein needed for efficient virus assembly. Congenital rubella may lead to a similar disease.

Progressive multifocal leukoencephalopathy (PMLE) is a chronic degenerative disease in patients who are immunologically compromised, for example those with lymphoma, leukaemia and transplants. A human polyoma virus 'JC' can be isolated from the oligodendrocytes in the affected areas of the brain. Silent infection with the virus is common in the population. PMLE is due either to a primary infection in someone who cannot respond, or to reactivation of a latent infection. AIDS patients suffer from dementia later in the illness and HIV can be demonstrated in brain tissue.

(b) *Unconventional agents*

Aetiology. The nature of these agents is unknown. The transmissible agent is small, associated with cell membranes and relatively resistant to inactivation. Clinically they produce progressive degenerative diseases of the CNS. The first demonstration of these in man was in Papua, New Guinea, where epidemiological evidence indicated that cannibalism or the preparation of the dead was important in transmission of the disease agent. It was transmitted to primates after a long incubation period and can persist in cell cultures from infected animals. No immune response can be detected and the disease has died out as cannibalism has been abolished. However, a parallel disease is Creutzfeld-Jakob disease, also rare, but recognised throughout the world. Transmission of this disease has been associated with neurosurgery and corneal grafting and the therapeutic use of human growth hormone prepared from cadavers. Agents with similar properties causing diseases with comparable features have been described in sheep and mink. The diseases are known as scrapie and Aleutian disease. In particular, scrapie has been intensively studied, as it can be adapted to the mouse and the incubation period reduced to a few months. From these studies evidence has been produced to suggest that slender 3–4 nm filamentous structures, or prions, can be found in extracts of infected tissues. Infectivity is associated with these extracts.

FURTHER READING

Bockman J M, Kingsbury D T, McKiney M P, Bendheim P E, Prusiner S B 1985 Creutzfeld-Jakob disease prion proteins in human brains. New England Journal of Medicine 312: 73–78

Christie A B 1980 Infectious diseases: epidemiology and clinical practice. Churchill Livingstone, Edinburgh

Davies P A 1977 Neonatal bacterial meningitis. British Journal of Hospital Medicine 19: 425–434

Duguid J P, Marmion B P, Swain R H A 1978 Mackie and McCartney: Medical microbiology. Churchill Livingstone. Edinburgh

Longson M, Bailey A S, Klapper P 1980 Herpes encephalitis. In: Waterson A P (ed) Recent advances in clinical virology. Churchill Livingstone, Edinburgh 2: pp 147–157

14

Infections of the locomotor system

Defences against infection

In the joint these include the synovial membrane and its cells which include specialised macrophages with a highly phagocytic action. Synovial fluid contains a few mononuclear cells, some complement and lysozyme.

Potential pathogens (see Fig. 14.1)

Acute septic arthritis

Bacterial aetiology

Bacteria commonly associated with this infection are *Staphylococcus aureus*, *Haemophilus influenzae*, beta-haemolytic streptococci and neisseriae, together with rarer causes such as *Brucella* spp, *Salmonella* spp, meningococci and gonococci.

Pathogenesis and epidemiology

Infection can occur by: (1) penetrating injury through the joint capsule; (2) secondary direct spread from a nearby focus of infection such as osteomyelitis; and (3) haematogenous spread to the capillary network in the synovial membrane and thence to the joint. The knee joints are most frequently affected, then the hip and ankle.

Sources of septic arthritis are numerous and include sepsis of the skin, nasopharynx, sinuses, lung, peritoneum and genital tract. Acute septic arthritis occurs most commonly in children. In an infant under 6 months of age the most likely organisms are *Staphylococcus aureus* or Gram-negative organisms of faecal origin; between 6 months and 2 years of age *Staphylococcus aureus* and *Haemophilus influenzae* are predominant, and over 2 years of age *Staphylococcus aureus* is again the major cause.

INFECTIONS OF THE LOCOMOTOR SYSTEM 217

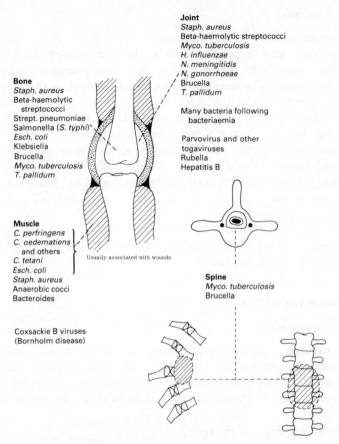

Fig. 14.1 Potential pathogens of muscle, bone and joint

Laboratory diagnosis

Blood cultures must be taken and synovial fluid should also be cultured. A primary site should be looked for and the appropriate material cultured. Synovial fluid appears turbid and creamy if infected, and will have a great increase in neutrophil polymorphs. (Exudate in rheumatoid arthritis looks turbid but no bacteria can be cultured).

Treatment

This will be dictated by the organism isolated and the results of sensitivity testing.

Viral arthritis

This may arise by two mechanisms:

(a) Virus may infect the synovia; the best example of this is rubella virus, both the wild type and some attenuated rubella vaccine strains. The transient arthritis occurs in the small joints in adults in natural infection. Adult females are most likely to develop arthralgia or arthritis after vaccination and, in a few cases, this may become chronic. Other togaviruses regularly infect joints. The human parvovirus has also been associated with arthritis especially in adult females. The arthritis usually affects the joints of the hands although wrists and knees can also be involved. The condition lasts for only a few weeks but a few patients have suffered a more protracted disease.

(b) Immune complexes may be formed at some stage in hepatitis B, and some togavirus infections. A macular rash often accompanies the joint involvement.

Acute osteomyelitis

Bacterial aetiology

Predominant bacteria are *Staphylococcus aureus*, beta-haemolytic streptococci, *Streptococcus pneumoniae* and Gram-negative bacilli, including *Haemophilus influenzae*, *Escherichia coli*, *Salmonella* spp and anaerobes. Tuberculosis and syphilis of bone can also occur.

Pathogenesis and epidemiology

This infection can occur in any bone and results from haematogenous spread from a septic focus (perhaps quite minor) elsewhere in the body. The type of organism depends on the source. In general terms coliforms are likely to spread from the urinary or intestinal tracts, whereas Gram-positive bacteria (mainly staphylococci, but also streptococci) may spread from the skin and upper respiratory tract. Anaerobes may originate in the intestinal tract, sinuses or teeth. The femur is most frequently affected, followed by the tibia, then the foot bones, humerus, fibula and pelvis. It is most common under the age of 12 and affects males more often than females.

Laboratory diagnosis

Blood cultures must be taken and if pus can be obtained this should also be processed. As with septic arthritis the primary site should be sought and material cultured. As over 90% of cases are caused by *Staphylococcus aureus*, flucloxacillin or another antistaphylococcal agent is the treatment of choice, although in the preschool child an anti-haemophilus antibiotic, such as amoxycillin or chloramphenicol, should be considered.

Viral myositis

Coxsackie B viruses are the most frequently identified causes of myositis. If this involves the intercostal muscles, a pleuritic-type pain results. This is known as Bornholm disease, epidemic myalgia or pleurodynia. Many togaviruses infect muscle, but this is often accompanied by rash. The extensive myalgia associated with influenza is similar to that seen in the prodromal phase of many acute infections.

Gas gangrene

Bacterial aetiology

Clostridium perfringens is the main organism involved. Others include *Clostridium oedematiens (novyi), septicum, histolyticum* and the *bifermentans-sordellii* group.

Pathogenesis and epidemiology

Gas gangrene is caused as the result of toxin production by clostridia and occurs when there is damaged or devitalised tissue which will provide the required conditions for anaerobic growth. *Clostridium perfringens* and other clostridia can normally be isolated from human and animal faeces and their spores are ubiquitous in nature. They can be isolated from the skin of the perineum and thighs and when gas gangrene results from surgical intervention in these areas it is usually an endogenous infection. This type of infection is seen in older patients following mid-thigh amputations for ischaemic disease of the lower limbs. Three types of wound infection are recognised: (1) *simple contamination* of wounds where the organisms do no harm, the commonest type of presentation; (2) *Clostridial cellulitis*, an acute, spreading infection of the subcutaneous tissues; and (3) *Clostridial myositis*, or true gas gangrene in which muscles are invaded; septicaemia and toxaemia are common.

Laboratory diagnosis

Gram-stained films demonstrate large Gram-positive bacilli and anaerobic blood agar cultures yield heavy growths of clostridia. Toxin production should be investigated.

Prophylaxis

Surgical removal of devitalised tissue is the basis of prophylaxis. When surgical procedures are to be carried out in the area of the thigh, perineum and buttocks, particularly in the elderly, prophylactic penicillin should be administered.

Treatment

Removal of devitalised tissue is again required as well as large doses of antibiotic, preferably penicillin. Hyperbaric oxygen therapy can be given in addition to polyvalent antitoxic serum.

FURTHER READING

Geddes A M 1981 Bone and joint infections. Medicine International 5: 186–188
Scott J T 1975 The analysis of joint fluids. British Journal of Hospital Medicine 14: 653–655
White D G, Woolf A D, Mortimer P P, Cohen B J, Blake D R, Bacon P A 1985 Human parvovirus arthropathy. Lancet i: 419–421

15

Infections of wounds, burns, skin and eye

Normal flora of skin

What constitutes 'normal' and what constitutes 'transient' flora is frequently difficult to establish. On the body average of two square metres of skin the various areas support different densities and species of organisms. The factors that determine selective localisation are far from clear. Different methods of sampling such as tape-stripping, scraping the *stratum corneum*, rubbing the skin with swabs and taking nutrient agar block impressions of the skin, can produce great differences in the composition and number of organisms. Three main factors determine the ecology of skin bacteria: (1) the micro climate and environment including temperature and humidity; (2) the effect of free fatty acids, other bacterial inhibitors and the low hydration of the *stratum corneum* in preventing colonisation by transient bacteria; and (3) maintenance of equilibrium of the resident flora by products of skin secretions.

The axillae, perineum and the scalp, areas of high humidity are densely colonised; the majority of the organisms reside in the superficial layers of the epidermal *stratum corneum*, although some 20% of organisms reside in the deep parts of the follicular canals which are the reservoirs that re-establish the skin flora when the surface has been cleared. All people disperse bacteria on particles of desquamated skin to a greater or lesser degree but some are described as 'dangerous shedders' or 'dispersers', particularly of *Staphylococcus aureus*, and are important from the point of view of cross infection. Except for the areas mentioned the general density of bacteria on the skin is not great.

The most common bacteria of the skin are as follows: coagulase-negative and coagulase-positive staphylococci; *Sarcina* spp; lipophilic, non-lipophilic and anaerobic diphtheroids; Gram-negative bacilli (in children) and various streptococci. Nutrients for these bacteria are derived from: (1) apocrine and eccrine sweat; (2) the *stratum corneum* whose water-soluble fraction contains amino acids and peptides; and (3) fatty acids from the sebaceous glands that form the main part of skin surface lipids. Water is of prime importance and the most densely populated areas are high in humidity; dry skin supports far fewer bacteria than moist skin.

The flora of the *scalp* (hair as well as skin) differs from that of other areas because the hair produces greater humidity and higher temperatures; it has in addition a rich sebaceous gland activity. *Staphylococcus aureus*, viridans streptococci, actinomycetes and coliform organisms can all be isolated from hair.

Defences against infection

These include the intact skin, bacteriocins, lysozyme in sweat and sebum, IgA and other inhibitors. The free fatty acids derived from hydrolysis of sebum triglycerides may be bactericidal or bacteriostatic, although in some instances they can also be stimulatory by providing nutritional factors.

Potential pathogens (see Table 15.1)

Table 15.1 Potential bacterial, fungal and protozoal pathogens of skin

Bacteria	
Staphylococcus aureus	Impetigo, Ritter's Disease (toxic epidermal necrolysis), folliculitis, furunculosis, carbuncle, ecthyma, sycosis barbae
Beta-haemolytic streptococcus	Impetigo, erysipelas, cellulitis, pemphigus
Erysipelothrix rhusiopathiae	Erysipeloid
Corynebacterium diphtheriae *Corynebacterium ulcerans*	} Cutaneous diphtheria
Corynebacterium acnes	Acne
Mycobacterium tuberculosis	Lupus vulgaris
Mycobacterium balnei	Swimming pool granuloma
Mycobacterium ulcerans *Mycobacterium marinum*	} Atypical cutaneous tuberculosis
Mycobacterium leprae	Leprosy
Bacillus anthracis	Anthrax (malignant pustule)
Haemophilus ducreyi	Chancroid
Treponemata	Syphilis, Yaws
Fungi	
Candida albicans	Cutaneous candidiasis
Microsporum canis *Microsporum audouinii* *Microsporum gypseum*	} Scalp ringworm (Tinea capitis)

Table 15.1 (contd)

Fungi (contd)
Trichophyton rubrum	Tinea pedis (athlete's foot)
Epidermophyton floccosum	
Trichophyton schoenleinii	Favus
Cryptococcus neoformans	Cryptococcosis
Actinomyces israelii	Actinomycosis
Nocardia madurae	Mycetoma (madura foot)

Protozoa
Leishmania tropica	Tropical sore
Leishmania brasiliensis	Espundia

Metazoa
Sarcoptes scabei	Scabies
Lice	Head, body, pubic region infestation
Fleas	

Infectious diseases
Many specific infectious diseases have skin manifestations that are important features of the diseases. These include:

Bacterial infections
Meningococcaemia (*Neisseria meningitidis*)
Scarlet fever (beta-haemolytic streptococcus)
Erythema nodosum

Infection of wounds

Bacterial aetiology

Many bacteria are associated with wound infection including *Staphylococcus aureus*, beta-haemolytic streptococci, *Escherichia coli*, *Proteus* spp, *Pseudomonas aeruginosa*, *Bacteroides* spp, *Clostridium perfringens*, *Clostridium tetani* and anaerobic cocci. Mixed infections are common.

Pathogenesis and epidemiology

Infection may occur following trauma, injections, bites and surgery. By far the most common wound infections are those occurring in hospitals after surgical procedures. Wound sepsis is the result of cross infection from human sources and from inanimate reservoirs. The skin and anterior nares are the most likely sources of staphylococci, the most frequent cause of wound infection, and spread occurs by the direct and indirect airborne routes and by contact. In a busy surgical ward bacterial counts per cm^3 of air (measured using slit samplers and sedimentation plates) rise during activities such as bed making and ward cleaning and fall when the ward is quiet. When a person moves about the numbers of airborne particles increase, partly because of the shedding of fresh particles from his body but also by raising particles that had fallen to the ground, and it is therefore very important to ensure that wards are not overcrowded and that hustle and bustle, particularly in the operating theatre, is kept to a minimum. Wound sepsis rates vary according to the type of surgical procedure performed, the size of the wound and the duration of the operation. Other factors such as the patient's age and general health, whether or not antibiotics are used and the efficiency of the post-operative care of the wound, all influence the sepsis rate.

Laboratory diagnosis

Pus is more satisfactory than a swab for this.

Prophylaxis

This depends on recognising the likely sources and reservoirs of infection both human and inanimate. Sources may be cases or carriers, staff or patient, and infection can be transmitted in the ward as well as in the theatre. Inanimate reservoirs include non-sterile equipment, bed linen and ward dust, and common reasons for infection are inadequate resources (for example, lack of dressing rooms), inexperienced staff, failure of aseptic techniques and failure to observe simple rules of hygiene.

Prophylaxis of wound infection depends on constant and careful surveillance of the ward and theatre environment, appraisal of sterilisation and aseptic techniques, nursing and medical procedures, and bacteriological monitoring of theatre and ward staff.

Management

When several cases of infection occur simultaneously or if the level of infection is generally at an unacceptably high level, simple epidemiological surveys and bacteriological searches for a source should be conducted. The most important act is to deal with the source if apparent. If this is a patient, isolation in a special cubicle or sideroom may be necessary and antibiotic treatment may be required. Staff who are infected should be kept away from the theatre and wards and treated. A *ward sepsis book* should be kept and close liaison with the laboratory should be maintained. An *Infection Control Officer*, often a Nursing Sister, is invaluable in such studies. Data required in the book include dates of admission, operation and onset of sepsis, type of operation and staff involved, theatre and ward involved, organisms isolated and antibiograms obtained. Such information may well pinpoint a particular individual, a break-down in some procedure, inadequate antiseptic and aseptic measures or faulty techniques. Phage typing, for example of *Staphylococcus aureus*, will indicate whether an epidemic strain is prevalent or whether the outbreaks are due to a number of other factors that can be remedied and more strictly controlled in the future.

Infection of burns

Bacterial aetiology

Organisms that colonise burns are derived mainly from the skin, upper respiratory tract and intestine of humans and include staphylococci, streptococci, coliform organisms and corynebacteria, although environmental bacteria such as aerobic spore-bearing bacilli may also be involved. The most important pathogens are those usually acquired by cross infection such as *Pseudomonas aeruginosa*, *Staphylococcus aureus* and beta-haemolytic streptococci, groups A, B, C and G. Gram-negative bacilli are also involved, and the flora of burns is usually mixed. Anaerobes play little part in infective complications of burns. The virus herpes simplex can also infect burns.

Pathogenesis and epidemiology

Any burn not given specific prophylaxis will become colonised by bacteria within 24 hours. Colonisation may lead to local sepsis with either cellulitis in the tissues next to the burn, or septicaemia. The necrotic surface of a burn with its concomitant moisture is a good culture medium in which the bacteria can multiply, out of reach of the body's natural defences. The levels of immunoglobulin are decreased for a few days after burning.

Before separation of the slough Gram-negative bacilli tend to predominate, whereas Gram-positive organisms and fungi such as *Candida* are predominant later in the granulation tissue. The presence of beta-haemolytic streptococci is always serious as they usually prevent the take of skin grafts. Invasion by these is rare however; invasive infections are generally caused by *Pseudomonas aeruginosa*. Superficial burns are not greatly harmed by colonisation but this is not so with the deep dermal burn.

Laboratory diagnosis

Specimens of necrotic material and swabs are processed on nutrient and blood agar, aerobically and anaerobically and in cooked-meat broth. Blood cultures are essential.

Prophylaxis

Because of the frequent failure of chemotherapy and the high mortality rate in severe burns *prophylaxis* is of paramount importance. Primary excision of dead tissue and closure of the wound is beneficial in preventing infection and the application of *topical antimicrobial agents* has been shown to reduce the colonisation rates of bacteria and subsequent infection. Agents used include silver nitrate solution, chlorhexidine cream and silver sulphadiazine cream, although the use of the latter causes a great increase in resistance of Gram-negative bacilli to sulphonamides. Antibiotics may be given systemically but are not always effective. Prophylaxis with immunoglobulin may be of use where topical chemoprophylaxis is difficult.

Infections of the skin and subcutaneous tissues

Acne

Bacterial aetiology. *Corynebacterium acnes*, anaerobic diphtheroids, propionibacteria, coagulase-negative staphylococci and micrococci are associated with this condition. *Pityrosporum* spp may also be involved.

Pathogenesis. Diffusible products of certain organisms may produce acne and the role of free fatty acids may be of importance.

Treatment. Many forms have been attempted. The role of antibiotics is questionable; *in vitro* testing does not identify an effective antibiotic, although some success is claimed for tetracyclines.

Bacterial infections of the skin

Staphylococcal infections

Boils or furuncles are the most common lesions caused by *Staphy-*

lococcus aureus. These infections are circumscribed, with suppurative central necrosis which eventually discharges as a small slough with pus. A *blind boil* subsides without discharge of pus. A *carbuncle* is a large abscess, much more serious than a boil, that occurs at the back of the neck in the thick collagenous tissues. It is a laterally-burrowing subcutaneous infection in which there are multiple openings to the skin, and it occurs most commonly in diabetics. *Staphylococcus aureus* also causes *breast abscess* in nursing mothers, and other superficial lesions that include *sycosis barbae* (face), *blepharitis* (eyelid), *stye* (eyelash follicle) and *conjunctivitis*. (Conjunctivitis is also caused by *Haemophilus influenzae* and *Streptococcus pneumoniae*). A form of *ecthyma, bullous impetigo* and *toxic epidermal necrolysis* are all caused by *Staphylococcus aureus*.

Streptococcal infections

These are mostly caused by *Streptococcus pyogenes* and include *cellulitis*, a spreading infection of the subcutaneous tissues and *erysipelas*, a similar infection of the dermis. In the latter, lesions occur on the limbs and on the face where the typical butterfly rash (cheek-nose-cheek) is seen. Repeated attacks of this infection are possible and may be due to the manifestation of a hypersensitivity reaction to the streptococci rather than to direct infection. Streptococcal *ecthyma* and *impetigo* also occur. This form of impetigo is less common than staphylococcal and is clinically different in that the lesions are yellow, typically discrete and subsequently form crusts. *Scarlet fever* is also associated with streptococcal infection, usually of the throat.

Gram-negative infections

Gram-negative organisms are rare on healthy skin except in moist areas such as the groin and axilla. *Proteus* spp are common in the latter site and this is thought to be due to the use of underarm deodorants. Coliform organisms and *Bacteroides* spp may be isolated from abscesses, particularly in the area between the thighs and the waist, and may be related to preceding trauma.

Otitis externa

In acute otitis externa Gram-positive bacteria are most frequently involved, whereas in the chronic form a wide range of bacteria may be isolated.

Paronychia (infection of the nail fold)

The acute attack is most commonly caused by Gram-positive organisms, whereas chronic infection is caused by a variety of bacteria, including *Pseudomonas aeruginosa*, as well as *Candida* spp.

Miscellaneous infections

One of the commonest infections, *trichomycosis axillaris*, is caused by overgrowth of diphtheroids on the shaft of axillary hairs and can lead to sweat discoloration of clothing. *Erysipeloid* is an infection of the hands and fingers caused by *Erysipelothrix rhusiopathiae*, and is an occupational hazard for those who handle infected animals or fish. Other infections include *syphilis, anthrax* and *leprosy*. A maculo-papular rash is a feature of several rickettsial infections within the spotted fever group.

Virus infections of the skin

Some viruses, for example warts, herpes simplex and molluscum contagiosum produce lesions at the site of inoculation. In most

Table 15.2 Virus infections of the skin

Clinical presentation	Viruses associated
Diffuse, erythematous	Parvovirus
Macular or Maculo-papular	Measles Rubella Enteroviruses Epstein-Barr virus Hepatitis B (in prodrome) HIV Lassa fever Marburg disease Togaviruses (dengue)
Vesicular	Herpes simplex, types 1 and 2 (especially severe in eczematous skin) Varicella zoster Enteroviruses

Solid—papillomas
  —molluscum contagiosum
	Warts viruses
	Pox virus—molluscum contagiosum
	  —orf

N.B. Rubella, the enteroviruses, EBV, herpes simplex virus, and hepatitis B are discussed elsewhere.

other cases however, the skin is a target often infected during the viraemic phase of a disseminated infection such as rubella, measles or chickenpox. Table 15.2 lists the viruses commonly associated with each type of rash.

Variola (smallpox) is not included, as it has been eradicated. It caused a severe disease with a vesicular eruption. Similarly vaccinia, the live virus in the vaccine to protect against smallpox, is not used in the United Kingdom and has been omitted from the list of causes of vesicular skin rashes. *Monkey pox* virus can infect man, but only those in close contact with the natural host.

Fifth disease or erythema infectiosum

This disease of children has recently been attributed to infection with a human parvovirus. The child presents with a confluent maculopapular rash of the cheeks—hence the alternative name of 'slapped-cheek syndrome'. The rash may be accompanied by vague malaise and fever. It is usually transient but may recur over some weeks. Infection is common; 80% of the population have antibodies to the virus.

Measles

Aetiology

Measles virus, a member of the paramyxovirus group.

Pathogenesis and epidemiology

Virus is shed from the mouth and upper respiratory tract late in the incubation period and throughout the period of the rash. Koplik's spots are recognisable at the onset of fever, respiratory symptoms and signs, before the typical maculo-papular rash develops at 10–14 days. Subclinical infections are uncommon. Lymphadenopathy and splenomegaly, abdominal pain, conjunctivitis, laryngitis and bronchiolitis all occur. Initial infection is silent in the upper respiratory tract; virus spreads rapidly to local lymph nodes and thence to blood and to skin and mucous membranes within 9–10 days. The rash develops when virus spreads in lymphocytes from capillaries to skin: the localised T-cell-dependent responses contribute to the production of the rash. Mucosal damage and a temporary reduction in immune responses contribute to bacterial superinfection, causing, for example broncho-pneumonia and otitis media. The tuberculin skin test may become negative during the period of leukopaenia induced by the virus. The disease has a high mortality in the malnourished.

An encephalitis or encephalomyelitis may occur shortly after the rash. A very rare late complication is subacute sclerosing panencephalitis (SSPE).

Diagnosis

Respiratory secretions should be collected by aspiration or by swab and sent for virus isolation. Blood samples are also collected to demonstrate a rising antibody titre.

Control

Only one virus type is known and epidemics occur every year or two. Eradication of the disease is possible with effective use of the attenuated live vaccine. Unvaccinated children who are immunosuppressed can be given immunoglobulin if exposed to infection.

Green monkey disease (Marburg disease)

This is a severe disease that was first recognised in laboratory workers who had been in contact with a batch of African green monkeys and their tissues. Isolated cases have since been recognised in South Africa. A related virus *Ebola virus* caused epidemics in the Sudan and Zaire in 1976. The disease is severe with fever, rash and haemorrhage: secondary cases occurred in the first outbreak including a case where transmission may have occurred via semen 2 months after infection in the index case.

Diagnosis

This is a category 4 pathogen and isolation can only be attempted in designated laboratories.

Enteroviruses

Some echoviruses, 4, 6, 9 and 16 and Coxsackie A9, A16, A23 can cause a maculo-papular rash. The viruses enter via the gut or respiratory tract and localise in the skin. These are mild diseases and their diagnosis may be confused with rubella. Coxsackie A16 is the cause of vesicular lesions in hand, foot and mouth disease. Dengue virus types 1–4 causes epidemics of infection in many tropical and sub-tropical areas. The febrile illness, with muscle and joint pains may be accompanied by rash. Re-infection occurs and can be associated with rash and haemorrhage—the *dengue shock syndrome* which has a much higher mortality than the primary infection.

Varicella and zoster

Aetiology

Varicella zoster virus.

Pathogenesis and epidemiology

The primary infection, chickenpox, is a disease of childhood with an incubation period of 14–17 days. The rash evolves rapidly from macular to papular to the vesicular stage. The rash is maximal on the trunk, but the limbs, neck and head are also involved. Successive crops of lesions develop, although in some cases there are only a few scattered vesicles. In adults the disease is more severe, and a viral pneumonia is usually present.

Virus reaches the skin from the blood; the bases of the skin lesions contain giant cells whose nuclei show intranuclear inclusions. The vesicle fluid contains high titres of virus, although viral shedding from the mouth and upper respiratory tract is probably the main source of infectious virus. A high attack rate (70%) may be recorded in close, susceptible contacts.

Immunocompromised patients may die of an overwhelming infection. The virus is not eliminated on recovery from chickenpox but becomes latent in the sensory root ganglia where it may remain for many years. Eventually, due to waning immunity or to intercurrent disease, reactivation occurs. This presents as *shingles* or zoster.

Zoster

The initial symptom is pain in the distribution of a sensory nerve, usually on the trunk, but the head and neck can also be involved. This is due to reactivation and growth of the virus within the ganglion and along the nerve. Virus reaches the skin and the typical vesicular rash develops. The disease is unilateral and usually only one or two dermatomes are affected. The rash is painful and may take 1-3 weeks to evolve to the crusting stage. On the head, the ophthalmic division of the fifth cranial nerve may be involved and lesions form on the conjunctiva and cornea. The mouth and ear may be affected—the Ramsay-Hunt syndrome. Extension to the anterior horn of the spinal cord can occur.

Most at risk of zoster are the elderly and especially those with some underlying disease, for example advanced malignant disease or those on immunosuppressive therapy. In such patients zoster is common and may disseminate. The patient with zoster is infectious because the virus is shed from the skin. This results in chickenpox in susceptible contacts, with an attack rate of up to 15%.

As many as one fifth of patients continue to experience pain in the previously affected dermatomes.

Diagnosis

Vesicle fluid, swabs or scrapes of the bases of lesions should be submitted for direct examination by electronmicroscopy or immunofluorescence. Virus isolation is also attempted but the virus grows slowly. Clotted blood samples should be submitted for serology. A rise in antibody titre may be demonstrated in chickenpox. In zoster, a very rapid rise occurs within a few days of the appearance of the rash. Sensitive techniques are needed to determine the immune status of those compromised persons exposed to the risk of infection.

Treatment and control

Passive immunisation with convalescent zoster immunoglobulin should be given to immunocompromised patients exposed to infection. Antiviral therapy with topical application of idoxuridine has shown some beneficial effect. However, systemic treatment with oral or parenteral acyclovir is now the treatment of choice. A live vaccine has been developed and is under trial.

Warts

Aetiology

Human papilloma virus (HPV): more than 40 different viruses have been identified by DNA hybridisation studies.

Pathogenesis and epidemiology

These infectious tumours occur in skin and mucous membranes at any site and present very different appearances, for example the simple hand wart, the plantar wart, genital wart or laryngeal papilloma. In all, there is proliferation of epithelial cells and the outer surface may show hyperkeratosis. Warts arising on the genitalia may become very large and troublesome especially in immunosuppressed patients. There are numerous virus particles in keratinising lesions. Infection is acquired by the direct transfer of virus from wart to skin, usually with some mild trauma, or the virus may survive outside the body until inoculated, as in transmission via gymnastic apparatus and the surrounds of swimming pools. The incubation period may be several months.

Skin warts usually regress spontaneously although the virus may persist in the tissues.

Genital warts are often multiple and are spread by sexual contact, although there are relatively few virus particles in the lesions. *Laryngeal papillomas* in children may be acquired from maternal genital warts, infection occurring during birth.

Oral papillomas occur in adults and may recur.

The importance of the immune response in controlling warts is indicated by the extensive lesions that develop in immunosuppressed patients. Malignant change in these lesions is now recognised and the role of warts virus in this has been investigated. The first clue was the definition that HPV types 3 and 5 are associated with the rare familial skin wart, *epidermodysplasia verruciformis* (EV) in which malignant transformation can develop on skin areas exposed to sunlight. Type 5 virus especially is associated with these malignant lesions. It has also been associated with similar lesions arising in renal transplant recipients.

Diagnosis

By clinical examination.

Treatment

By cytotoxic therapy applied locally, or by excision.

Molluscum contagiosum

Aetiology

Poxvirus: molluscum contagiosum virus. The small fleshy lesions can occur anywhere on the skin of trunk, arms and around and on the eyelids and are usually single or in small clusters. The skin of the genitalia and nearby areas is another common site. The immunocompromised patient may develop more numerous lesions. Virus is shed from the core of the lesion. There is no treatment; the infection is self-limiting.

Diagnosis

Diagnosis is by EM of core of lesion.

Orf

Aetiology

Poxvirus—orf or contagious pustular dermatitis of sheep. This relatively rare disease presents as a raised solid lesion 1–2 cm in diameter, often with associated infection and lymphangitis. It is an occupational hazard for shepherds, veterinary and slaughter-house workers and butchers. The lesions are usually single and on the hands. There is much virus in the lesion, but man to man transfer is not recorded. There is no specific treatment.

Diagnosis

By EM of material scraped from the lesion, to show the characteristic particle with rounded ends and symmetrical surface tubules.

INFECTIONS OF THE EYE

Orbital cellulitis

Infection usually spreads from adjacent tissues such as teeth and sinuses and the organisms involved are generally those relevant to these areas.

The lacrimal (lachrymal) apparatus

The lacrimal gland can be infected by local spread of organisms from nearby sites. Acute inflammation of the gland (dacryoadenitis) may also be associated with mumps and gonorrhoea. The lacrimal canaliculi may be similarly infected.

Conjunctivitis

(a) Acute

If there is discharge which is purulent this usually indicates a bacterial infection. The organisms involved vary and include *Staphylococcus aureus*, *Haemophilus aegyptius* (Koch-Weeks bacillus), pneumococci, gonococci and meningococci. *Pseudomonas aeruginosa* may invade the conjunctiva after eye operations.

(b) Mucopurulent

This can be an early sign of ophthalmic herpes zoster and can be caused by *Chlamydia trachomatis*. Infection with serotypes A, B, C causes trachoma, a disease of the malnourished, living in poor conditions. The organism may be transferred directly on fingers or by flies; re-infection is common. The disease starts early in life, and may cause blindness due to scarring and vascularisation of the cornea. Infection with serotypes D-K is associated with the genital tract, but a benign conjunctivitis may accompany this. Infection from swimming pools has been recorded.

(c) Nonpurulent

This is commonly associated with viruses, notably adenoviruses types 3, 7 and 14. The conjunctivitis is often accompanied by other manifestations of adenovirus infection such as pharyngitis (pharyngo-conjunctival fever), lymphadenopathy or a vague febrile illness. Adenovirus types 7 and 8 cause a kerato-conjunctivities ('shipyard eye') in persons who have repeated corneal abrasion, due to dust, grit or metal particles. Herpes simplex virus may also cause primary and recurrent episodes of keratoconjunctivitis. In most cases of nonpurulent conjunctivitis, however, no microbial causes is identified.

(d) Neonatal

Organisms that produce conjunctivitis in adults can also do so in infants but the latter are additionally liable to infection from the mother's genital tract during labour.'*Ophthalmia neonatorum*' is defined in Britain as a purulent discharge from the eyes within 3 weeks of birth. Gonococci, chlamydiae, staphylococci, streptococci (group B), pseudomonads and coliforms may be involved. *Staphy-*

lococcus aureus and herpes simplex can also be acquired from attendants at the birth, or other hospital personnel.

Keratitis

Any infective causes of conjunctivitis can spread deeper to involve the cornea, causing ulcers or a spreading keratitis. Viruses such as herpes simplex, herpes zoster and vaccinia, bacteria such as pneumococci, gonococci and pseudomonads and fungi are associated with keratitis.

Herpetic keratitis

This may rarely present as a primary infection or, by extension, from infection of the skin of the face and orbit; however it is usually seen as a recurring problem. Virus is reactivated from the trigeminal ganglion, and reaches the cornea via the tears. Corneal ulceration produces a branching pattern to form a dendritic ulcer. The disease is usually unilateral, but frequent recrudescences lead to corneal opacity, which may need corneal grafting to improve sight. This does not remove the source of the virus and the problem may recur. The use of steroids locally may encourage HSV infection. Acyclovir can be given locally.

Varicella zoster

Shingles of the ophthalmic division of the fifth cranial nerve may involve the cornea.

Cataract

Congenital rubella.

Chorioretinitis

Congenital rubella, CMV reactivation in the severely immunocompromised and congenital infection with *Toxoplasma gondii* can cause this.

Diagnosis

Smears can be examined directly for bacteria, and by immunofluorescence for *Chlamydia trachomatis* and herpes simplex virus. Swabs for bacterial isolation should be sent to the laboratory without delay or inoculated at the bedside. Swabs or scrapes for virus isolation should be placed in virus transport medium. Note that a separate transport medium is essential for *Chlamydia trachomatis* isolation. Paired serum samples can be submitted for serological tests.

Endophthalmitis and panophthalmitis

Infections of the inner eye derive from infections of the outer eye, penetration of foreign bodies and infection during surgical operations.

FURTHER READING

Anderson M J, Pattison J R 1984 The human parvovirus. Archives of Virology 82: 137-148
Ayliffe G A J, Collins B J, Taylor L 1982 Hospital-acquired infection. Wright P S G, Bristol
Gibson G L 1974 Infection in Hospital. Churchill Livingstone, Edinburgh
Lowbury E J L 1976 Prophylaxis and treatment for infection of burns. British Journal of Hospital Medicine 15: 566-572
Lowbury E J L, Ayliffe G A J 1982 Airborne infection in hospital. Journal of Hospital Infection 3: 217-240
Lowbury E J L, Ayllife G A J, Geddes A M, Williams J D 1982 Control of hospital infection: a practical handbook. 2nd edn. Chapman and Hall, London
Marks R, Samman P D 1977 Dermatology. Heinemann, London
Meers P D, Ayliffe G A J, Emmerson A M et al 1980 Report on the National Survey of Infection in Hospitals. Journal of Hospital Infection 2: 13-17
Seal D V 1983 The role of antiseptics and disinfectants in the control of nosocomial infection. British Journal of Clinical Practice suppl, 25: 46-54
Hansen H, Gissmann L, Schlehofer J R 1984 Viruses in the etiology of human genital cancer. Progress in Medical Virology 30: 170-186

16

Miscellaneous bacterial infections

Actinomycosis

Bacterial aetiology

In man the principal infecting organism is *Actinomyces israelii*. *Actinomyces naeslundii* may also be involved and *Actinomyces bovis* produces actinomycosis in cattle.

Pathogenesis and epidemiology

Actinomycosis can occur in the lung and ileo-caecal region but by far the most common areas are the face and neck. It is an uncommon, chronic suppurative and granulomatous infection with sinus formation and discharge of pus containing '*sulphur granules*' that are colonies of *Actinomyces*. *Actinomyces israelii* is a normal inhabitant of the mouth. Infection of the cervicofacial region occurs by local invasion and may follow trauma such as fracture of the jaw and tooth extraction, or it may be due to the presence of teeth with gangrenous pulps.

Laboratory diagnosis

Pus must be examined for the presence of granules which if present should be crushed, and a Gram film made. Granules consist of Gram-positive branching filaments with Gram-negative clubbed ends, which on tissue section can be seen to radiate from a central mass of mycelium. Cultures are made on blood agar incubated anaerobically for 4–7 days. Further fermentation tests are required for identification.

Treatment

Penicillin is the most appropriate drug in addition to necessary surgical measures. A six-week course of treatment is recommended. Tetracyclines are alternative drugs.

Anthrax

Bacterial aetiology

Bacillus anthracis.

Pathogenesis and epidemiology

Anthrax, a zoonosis, is a septicaemic infection of herbivora, mainly sheep and cattle, although virtually all animals are susceptible to some degree. The disease is uncommon in the United Kingdom and outbreaks that occur, for example in zoos, are due to ingestion of spores in contaminated foodstuffs. Human anthrax is uncommon and occurs by contact or inhalation in people dealing with animals or products of animals that have died from the disease, for example shaving brushes, wool, skins, bones and bone meal, hair and bristles. Two forms of human anthrax occur, cutaneous infection, *malignant pustule*, and very rarely a haemorrhagic infection of the lung, *woolsorters' disease.*

Laboratory diagnosis

Vesicular fluid or swabs from a suspected malignant pustule should be examined for large Gram-positive bacilli; these are virtually diagnostic. Various cultural features such as characteristic colonies and non-motility distinguish anthrax bacilli from other sporing bacilli, but pathogenicity tests on mice or guinea pigs are essential for confirmation of identity.

Prophylaxis

Animals may be immunised against anthrax. Infected animals must be killed and cremated or buried deeply. Rigid control of importation of hides and protection of the hands and other exposed skin surfaces of workers have reduced the incidence of the disease.

Treatment

Penicillin is the drug of choice but may not be effective in the serious respiratory form.

Erysipeloid

Caused by *Erysipelothrix rhusiopathiae* this is a lesion usually of the fingers that resembles erysipelas. The organisms are found in birds, fish and animals and infection is an occupational disease of workers handling meat and fish. Penicillin is the treatment of choice.

Leprosy

Bacterial aetiology

Mycobacterium leprae.

Pathogenesis and epidemiology

The organism is a human parasite that cannot be made to infect laboratory animals by conventional methods, although experimental infection of the mouse foot pad is possible. The infectivity of the disease is low and requires long and close contact; it is probably contracted both through skin abrasions and also by inhalation of infected dust. It is now mainly confined to tropical countries such as Ethiopia.

Leprosy is a chronic granulomatous infection. In the *lepromatous* and *nodular* types granulomata form in the skin, mucosae and various organs. In the *tuberculoid* type there is infiltration of certain motor and sensory nerves.

Laboratory diagnosis

This depends on demonstrating acid-fast bacilli in smears and tissue scrapings. The organisms resemble *Mycobacterium tuberculosis* and bacteria are commonly seen inside mononuclear lepra cells. The organisms cannot be grown on artificial media.

Prophylaxis

Vaccination has been recommended particularly against tuberculoid leprosy, but further studies are required to confirm the usefulness of this. Health education is also important.

Treatment

Oral sulphones, clofazimine or rifampicin may be used.

Leptospiral infections

(a) Weil's disease

Bacterial aetiology

Leptospira icterohaemorrhagiae.

Pathogenesis and epidemiology

The organisms which produce leptospirosis, a zoonosis, are normally carried in the kidneys of the brown rat. They cause no harm to the rats, but are excreted in their urine usually for life; they can live for days in stagnant water, mud and moist earth, and infection occurs through abrasions in the skin and mucous membranes. Agricultural workers, miners, fish handlers and sewer workers are all liable to infection but leptospirosis can also be caused by drinking infected water or bathing in canals, ponds or rivers. Leptospires multiply in the bloodstream and spread to the liver, kidneys and lungs, which accounts for the *protean manifestations* of the disease. They have a particular predilection for the kidney cortex where they colonise the convoluted tubules and remain there for some time.

(b) Canicola fever

Bacterial aetiology

Leptospira canicola.

Pathogenesis and epidemiology

Another leptospiral serotype causes this disease which often presents as meningitis, with particularly severe headache and conjunctivitis. Puppies are prone to canicola fever. Children contract it from the infected excreta of a sick puppy, particularly one with nephritis. Pigs are also commonly infected.

Laboratory diagnosis of leptospiral infections

After an incubation period of 5–12 days leptospires may be seen in the blood by dark-ground microscopy and in the second week of the infection they may be seen in the urine. Guinea-pig inoculation of body fluids may be useful, and special media for blood and urine culture are available. Serological tests are useful as antibodies are detectable from the end of the first week; agglutination tests and complement fixation tests are widely used.

Prophylaxis

Education of dog owners is important and workers at risk should wear suitable protective clothing. Puppies should be immunised. Prophylactic vaccination is also given to certain groups of workers where the risk is high.

Treatment

Leptospirosis may be treated by penicillin which is the drug of choice. Tetracyclines are alternative agents, although in general antibiotics are only effective in influencing the course of the disease if they are given in the first week of the illness.

Listeriosis

This is a rare infection caused by *Listeria monocytogenes* that is isolated from wild and domestic animals. Septicaemia and meningitis can be caused in neonates compromised patients and abortion or premature stillbirth may occur in women who have chronic infection of the genital organs. Ampicillin or chloramphenicol are used in treatment.

Plague

Bacterial aetiology

Yersinia pestis.

Pathogenesis and epidemiology

The plague bacillus is a parasite of rodents (particularly rats) and their fleas spread infection to other animals. Infected rats living around human habitation are commonly responsible for outbreaks of human plague and the infection is spread to man by infected rat fleas. Person-to-person infection is rare except from a patient with *pneumonic plague* who may disperse infected droplets in the environment. In *bubonic plague* lymph nodes are affected, in particular those draining the area of the flea bite. Glands (buboes) swell in the groin, axilla or neck depending on the site of the bite. Septicaemia and death may thereafter occur. In *pneumonic plague* the patient suffers from a severe haemorrhagic bronchopneumonia. The virulence of the plague bacillus relates to factors that protect it from phagocytosis, such as capsular and somatic antigens.

Laboratory diagnosis

Diagnosis is confirmed by isolation of bacilli from buboes or sputum. The bacilli show characteristic bipolar staining with methylene blue. Blood cultures may yield the organism and should be taken in suspected cases. Isolation on blood agar is facilitated by anaerobic conditions at 27°C. Inoculation of pathological material on to the nasal mucosa of guinea pigs or rats may yield the pure cultures necessary for biochemical identification.

Prophylaxis

This essentially means control over wild and domestic rat populations and particularly fleas if epidemics threaten. Vaccination can provide immunity of short duration.

Treatment

Large doses of tetracycline are recommended for both bubonic and pneumonic plague.

Rat-bite fever

Bacterial aetiology

Spirillum minus and *Streptobacillus moniliformis* produce diseases that are clinically indistinguishable. Both organisms can be isolated from the respiratory tract of rats.

Pathogenesis and epidemiology

Rat-bite fever is a relapsing febrile illness with local inflammation at the site of the bite, enlargement of local lymph glands and a macular skin eruption. A multiple arteritis may occur. *Spirillum minus* infection occurs in Japan and the Far East whereas *Streptobacillus moniliformis* infection occurs in the Americas.

Spirillum minus can be demonstrated by dark-ground illumination in fresh preparations of lesions or lymph glands and also in the blood of mice or guinea pigs 10 days after injection. The organisms can be stained by Leishman's stain or by ordinary aniline dyes, but as they cannot be cultured on inanimate media some workers class them as spirochaetes. Spirilla differ from these however in that they are rigid and owe their motility to polar flagella.

Streptobacillus moniliformis can be isolated from blood cultures and from fluid media that have a high content of blood, serum or ascitic fluid. They are highly pleomorphic and L forms occur.

Treatment

Both forms respond well to penicillin and chloramphenicol and tetracycline is active against L forms of *Streptobacillus moniliformis*.

Pasteurella multocida infection

Pasteurella multocida (Pasteurella septica): This is found in domestic animals, particularly dogs and cats, and poultry. Man is infected by bites and scratches and respiratory infection and meningitis, as well as local sepsis, may be produced. The short Gram-negative bacillus may be isolated from wounds, blood, sputum or CSF. Benzyl penicillin is used in treatment.

Pyrexia of unknown origin

Known as PUO, this can be a feature of tuberculosis, infective endocarditis, intestinal and urinary tract infections as well as localised sepsis of virtually all body systems. EB virus, CMV, hepatitis and enteroviruses may be involved as well as chlamydiae, coxiella and protozoa. Atypical presentation is common and diagnosis difficult.

Relapsing fever

Bacterial aetiology

The causative organisms are *Borrelia recurrentis*, the infecting organism of European (louse-borne) relapsing fever, and *Borrelia duttonii*, the infecting organism of African (tick-borne) relapsing fever.

Pathogenesis and epidemiology

Louse-borne fever may be endemic or epidemic in conditions of cold and overcrowding. After lice become infected they cannot transmit infection for a few days; the multiplying spirochaetes then enter the coelomic cavity. Coelomic fluid of infected crushed lice or infected excreta may enter the human through abrasions. Bouts of fever lasting from 2–4 days are separated by intervals of around 1 week. Many relapses can occur and it is thought that each relapse is produced by fresh antigenic types of organisms. Wild rodents are the principal sources of infection with *Borrelia duttonii*. Human cases result from tick bites or contamination of wounds with infected secretions. The organisms persist throughout the life of the tick and are transferred to succeeding generations through the ova. Unlike louse-borne fever, tick-borne fever does not occur epidemically.

Laboratory diagnosis

This is made by the demonstration of borreliae in blood films stained by Giemsa's or Leishman's methods or dilute carbol fuchsin, particularly during pyrexial phases. Dark-ground microscopy of blood is also useful. Inoculation of rats produces considerable numbers of borreliae in their blood within 48 hours.

Treatment

Tetracyclines, penicillin and chloramphenicol are all effective in the treatment of these infections.

Tularaemia

Francisella tularensis infects rodents, dogs and cats, particularly in America and Scandinavia, but not in the United Kingdom. Man is infected by scratches and bites. Lymphadenopathy and skin ulceration are caused and streptomycin is used in treatment.

Rickettsial infections

Rickettsiae are intracellular organisms and are responsible for some important diseases of man, such as the typhus fevers, spotted fevers, scrub typhus and trench fever. Q fever is caused by a related organism *Coxiella burneti* (see Ch. 6).

Classical or epidemic typhus is caused by *Rickettsia prowazeki* and is transmitted by the body louse. Features of the disease are severe headache, fever, aches and vomiting, followed by a rash. Large outbreaks occur when there is a breakdown in hygiene, as in war. The organisms grow in the endothelial cells of small blood vessels of many body systems. The mortality rate varies with age, from less than 40% in patients up to 40 years of age to more than 50% in those over 50.

Organisms can persist after the initial infection and this may cause a mild recrudescent illness called *Brill-Zinsser* disease.

Isolation of the organisms is difficult and dangerous and serological tests are the usual means of diagnosis.

The organisms are sensitive to tetracyclines and chloramphenicol. Vaccines have been developed, but the most effective control measure is the elimination of the body louse and the conditions which favour its spread.

Murine typhus is a sporadic, mild disease caused by *Rickettsia typhi*. The rat is the natural host and man is infected by the bite of the rat flea.

The *spotted fevers* are caused by a number of different rickettsiae in many parts of the world. Infection is endemic in a vertebrate host and is transmitted by ticks. *Scrub typhus* occurs in Asia, in the warm and wet regions; infection is caused by *Rickettsia tsutsugamushi* and is spread by mites with reservoirs in various mammals and birds.

Trench fever was first defined in the 1914–18 war but it also affected military personnel in the 1939–45 war. The organism responsible, originally named *Rickettsia quintana*, differs from the true rickettsiae in that it can be grown on cell-free media.

FURTHER READING

Christie A B 1980 Infectious diseases: epidemiology and clinical practice, 3rd edn. Churchill Livingstone, Edinburgh
Duguid J P, Marmion B P, Swain R H A 1978 Medical microbiology. Churchill Livingstone, Edinburgh
Lamb R 1973 Anthrax. British Medical Journal I: 157–159
Lawson J H 1971 Leptospirosis. British Journal of Hospital Medicine 5: 357–364

Index

Abscesses, 227
  breast, 192
  cerebral, 209–10
  dental, 139–40, 142
  lung, 110, 116
  pelvic, 160
  perinephric, 175
  pyogenic liver, 159
  subphrenic, 159–60
Acne, 226
Acquired immunity, 36–42
acquired immune deficiency syndrome
  see AIDS
Actinomyces spp, 83, 133, 134, 135, 136, 137, 139, 142, 238
Actinomycosis, 62, 238–9
Acute
  epiglottitis, 104–5
  glomerulonephritis (AGN), 173–5
  laryngo-tracheo-bronchitis, 104–5
  meningitis, 199
  osteomyelitis, 218–19
  pericarditis, 125
  pyogenic meningitis, 200–3
  septic arthritis, 216–18
  ulceromembranous gingivitis, 140–1
Acyclovir (acycloguanosine), 94–5
Adenine arabinoside, 94
Adenoviruses, 103, 105, 130, 148, 176, 235
Adherence, of organisms, 23–4
Aggressins, 21–3
AIDS (acquired immune deficiency syndrome), 11, 55, 93, 95, 127, 128–30, 144, 181
  and infectious agents, 128, 130,
Airborne spread, 6–7, 10
Alcohols as disinfectants, 51
Aldehydes, 47, 51
Aleutian disease, 215
Amantadine, 95, 109
Amoebic dysentery, 155
Amikacin, 88
Aminoglycosides, 87–9

Amoxycillin, 85
Ampholytes, 54
Ampicillins, 84–5
Anaemias, 126
Animal pathogenicity tests, 70
Animals, spread infection, 3, 10
  zoonoses, 3, 158, 239, 241
  see also Specific diseases
Anthrax, 62, 228, 239
Antibiograms, 70
Antibiotics, 74–92
  see also Antimicrobial agents
Antibodies, 32–3, 35, 37, 38, 66
  detection, 71–2
Antigen detection, 68, 69–70
Antigenic drift and shift, 108
Antikidney antibody, 174
Antimicrobial agents, 74–96
  antiviral agents, 32, 93–6
  clinically important, 81–92
  laboratory control of therapy, 80–1
  modes of action, 77–8
  principles of use, 74–6
  combinations, 75–6
  resistance, 78–80
Antimicrobial antibody actions, 34
Antisepsis, 48
Antiviral agents, 93–6
  interferons, 32, 95–6
Arachnia, 137
Arthritis, 63, 216–18
Arthropods, vectors of disease, 10
Aspergillosis, 117
Astroviruses, 148
Atypical pneumonia, 111
Augmentin, 85
Autoclave, 44–6
Azlocillin, 85

Bacillary dysentery, 154–5
Bacillus spp, 133
  anthracis, 14, 83, 239
  cereus, 148, 150
  stearothermophilus, 46

247

Bacitracin, 91
Bacteriaemia, 17, 118
Bacterial pathogenicity
  adherence, 23-4
  aggressins, 21-3
  capsules, 21-2
  invasiveness, 24-8
  toxigenicity, 18-21
Bactericidal
  action, 74-5
  secretions, 29-30
Bacteriogenic shock, 20
Bacteriostatic action, 74-5
Bacteriuria, asymptomatic, 169-73
Bacteroides spp, 116, 118, 120, 121, 145, 160, 177, 180, 192, 209, 223, 227
  and drugs, 89, 90, 92
  in mouth cavity, 133, 134, 136, 137, 139
  pathogenicity, 20, 23
  gingivalis, 142
  melaninogenicus, 141
Balanitis, 181
Balantidiasis, 157
Bartholinitis, 178
BCG, 39, 114-15
Benzimadazoles, 95
Benzyl penicillin, 83
Bifidobacterium, 145
Biopsy specimens, 61, 67
BK virus, 176
Blood
  changes, in infection, 26-7
  common isolates, 118
  for culture, 58-9, 120-1
  spread of infections, 16-17, 55-6, 129, 163-4, 165-6
  vessels, damaged, 126
Boil, staphylococcal, 15, 226-7
Bordetella pertussis, 24, 90, 106-7
Bornholm disease, 219
Borrelia
  duttonii, 244, 245
  recurrentis, 244-5
  vincenti, 141
Botulism, 210-11
Bowie-Dick tape, 45
Breast abscess, 192
Brill-Zinsser disease, 246
Bronchiolitis, 109
Bronchitis, 105
Bronchopneumonia, 109-10, 116
Browne's tubes, 45
Brucella spp, 1, 8, 17, 18, 31, 35, 118, 158

Brucellosis, 63, 158-9
Bubonic plague, 15, 242-3
Burn infections, 225-6

Caliciviruses, 147-8
Campylobacter spp, 137, 141, 148, 153-4, 196
Candidiasis, 128, 227, 228
  genital, 182, 183
  oral, 130, 140, 197
Canicola fever, 241-2
Capsid, viral, 24
Capsule, bacterial, 21-2
Carbenicillins, 85-6
Carbuncle, 227
Carcinoma, of cervix, 189
Cardiovascular infections, 26, 118-31
  defences, 118
  potential pathogens, 119
  predisposing factors, 119
  viral infections, 127-31
  see also Specific diseases and disorders
Carfecillin, 86
Caries, 135-7
Carriers, of infection, 2-3
Cataract, 236
Cell cultures, 70-1
Cell-mediated immunity, 34-5, 36
Cellulitis, 227
  orbital, 234
  pelvic, 180-1
Central nervous system infections, 199-215
  chronic virus infections, 212, 214-15
  defences, 199
  potential pathogens, 199
  spread of infections, 16, 18
  see also Specific diseases and disorders
Cephalosporins, 86-7
Cerebral
  abscess, 209-10
  toxoplasmosis, 309
Cerebrospinal fluid (CSF)
  in meningitis, 65, 200, 202, 204
  specimens, 59
  spread of infections, 16, 204
Cervical
  carcinoma, 189
  swab/scrape, 59, 62
Cervicitis, 11, 179-80, 187, 188
Chancre, 185
Chancroid, 180, 182, 183
Chemical disinfectants, 51-6
Chemotaxis, 30-1

INDEX 249

Chemotherapy see Antimicrobial agents
Chicken pox, 33, 197, 231–2
Chick-Martin test, 50
Childbed fever, 191–2
Chlamydia spp, 57, 70–1, 89, 90, 112, 121
  psittaci, 112
  trachomatis, 103, 176, 180, 181, 182, 186–7, 195, 196, 235
Chlorhexidine, 49, 50, 51–2, 54, 55
Chlorine, 52
Cholera, 5, 8, 155–7
  infantum, 152
Chorioamnionitis, 191
Chorioretinitis, 236
Clindamycin, 91
Clostridium spp, 83, 118, 145, 219–20
  botulinum, 19, 210–11
  difficile, 157
  perfringens, 3, 19, 22, 23, 77, 148, 149, 219–20, 223
  tetani, 3, 9, 14, 19, 77, 211–12, 223
Cloxacillins, 84
Coagulase, 22
Cocci, anaerobic, 99, 116, 118, 120, 137, 142, 145, 177, 192, 209, 223
Colistin, 91
Colitis, 153, 157
Collagenase, 23
Colorado tick fever, 126
Common cold, 99
Complement fixation
Congenital infections, 63, 192–5, 236
  see also Pregnancy
Contagious pustular dermatitis, 234
Control of infections, 11–12
Coronaviruses, 98, 99, 148
Corynebacterium
  acnes, 226
  diphtheriae, 100–2
Cotrimoxazole, 82–3
Counterimmunoelectrophoresis, 68, 69
Cowdry type A inclusions, 21
Coxiella burneti, 8, 89, 111, 120, 121, 122, 162, 245
Coxsackieviruses, 99, 125, 197, 203–4, 219, 230–1
Creutyfeld-Jakob disease (see central nervous system) 212, 214
Croup, 104–5
Cryptococcus neoformans, 130
Cryptosporidium
Culture, of organisms, 68–9, 70–1
Cystic fibrosis, 110
Cystitis, 169, 176, 187

Cytarabine (cytosine arabinoside), 94
Cytomegalovirus (CMV), 118, 127, 130, 143, 144, 161, 175, 236
  congenital infections, 11, 191, 193, 194–5
Cytoplasmic inclusions, 21

Dengue, 10, 231
Dental
  abscesses, 139–40, 142
  caries, 135–7
  plague, 134, 135–6, 141
Denture stomatitis, 140
Deoxyribonucleases (DNAases), 23
Diagnosis of infections, 57–73
  collection of specimens, 57–62
  laboratory methods, 67–73
  protocols of investigation, 62–6
Diarrhoea, 63, 157
  see also Gastrointestinal tract infections
Diffusion tests, 80, 81
Diguanides, 51–2, 53
Dilution tests, 80, 81
Diphtheria, 15, 39, 100–2
Diphtheroids, 177, 226, 227
Dip slides, 171
Disinfection, 47–56
  chemical disinfectants, 51–4
  efficiency testing, 50–1
  summary, 54–6
Disseminated intravascular coagulation, 20–1, 35
Donovania granulomatis, 182
Droplet spread, 6–7
Drug abuse, 10, 120, 129, 163
Drugs see Antimicrobial agents
Drum-stick bacilli, 211
Dysentery, 154–5

Ebola virus, 230
Echoviruses, 99, 203, 204, 230–1
Ecthyma, 227
Electrophoresis, 68, 69
ELISA (enzyme-linked immunoassay), 68, 69–70, 72, 112, 147
El Tor vibrio, 155, 156
Empyema, 116, 117
Encephalitis, 205–7
Encephalomyelitis, post-infectious, 207
Endemic infection, 4–5
Endocarditis, infective, 63–4, 120–2
Endogenous infections, 1–2
Endometritus, 11, 180
Endophthalmitis, 237

Endotoxins, 18, 19, 20–1
  endotoxic shock, 20
Entamoeba spp, 133, 155, 159
Enteric fever, 150–2
Enterocolitis, 153–4
Enterotoxins, 149, 150, 152, 154, 156, 157, 196
Enteroviruses, 103, 161, 197, 203–4, 230–1
Enzootic infections, 5
Epidemic infections, 5
Epidermodysplasia verruciformis (EV), 233
Epididymitis, 181
Epiglottitis, acute, 104–5
Epizootic infections, 5
Epstein-Barr virus (EBV), 14, 127–8, 144
Erysipelas, 227
Erysipeloid, 228, 240
Erythema infectiosum, 229
Erythrocytes, and virus infections, 126
Erythromycin, 90
Escherichia coli, 20, 24, 118, 191, 195, 196, 218, 223
  and drugs, 82, 83, 84, 88, 173
  in gastrointestinal tract, 145, 148, 152–3, 157
  in urinary tract infection, 1, 82, 168, 172, 173
Ethylene oxide, 47, 55
Exogenous infections, 2
Exotic infections, 5
Exotoxins, 18–19
Eye infections, 59, 234–7

Faecal specimens, 59
Fever, 26
Fibrinolysin, 22–23
Filtration, in sterilisation, 46–7
Flucloxacillin, 84
Fluorescent tests *see* Immunofluorescence
Fomites, 6, 9
Food, contaminated, 7, 8
Food poisoning, 64, 146, 148–50, 156, 210–11
Foreign bodies, and infective dose, 14
Formaldehyde, 47, 48, 51
Framycetin, 88
Francisella tularensis, 245
Fungal infections, 3, 117, 120, 222–3
  see also Candidiasis
Furuncle, 15, 226–7
Fusidic acid, 91

Fusobacterium spp, 102, 116, 134, 136, 137, 139, 141, 209

Gangrene, 64, 77, 219–20
Gas sterilisation, 47, 51, 55
Gastroenteritis, 146, 147–8
  infantile, 152–3, 196
Gastrointestinal infections, 145–67
  defences, 145
  liver infections, 161–7
  normal flora, 145
  potential pathogens, 145–6
  spread of infections, 8, 12
  see also Specific diseases and disorders
Genetic factors, 24–5, 29
Genital tract infections, 177–89
  defences, 177–8
  laboratory investigation, 64
  neonatal, 112, 235–6
  potential pathogens, 179
  sexually transmitted disease, 169, 180, 181–9
  spread of infections, 11
  therapy, 94, 95
  see also Specific diseases and disorders
Genito-urinary tuberculosis, 173
Gentamicin, 88
Giardiasis, 92, 154, 157
Gingivitis, 139, 140–2
Glandular fever, 127–8, 161
Glomerulonephritis, acute, 173–5
Glutaraldehyde, 48, 49, 51, 55
Gonococcal ophthalmia, 183
Gonorrhoea, 64, 83, 89, 180, 182, 183–4, 234
Granuloma inguinale, 182
Green Monkey Disease, 230
Guanidine, 95
Gummata, 185

Haematological changes, in infection, 26–7
Haemophilus spp, 89, 117, 118, 139, 177, 201, 203
  aegyptius, 235
  ducreyi, 182, 183
  influenzae, 16, 22, 82, 84, 90, 120, 201, 209, 216, 218, 227
  respiratory infections, 99, 104, 105, 107, 110
Halogens, 52
Heaf test, 115
Heat sterilisation, 42–6
Helminth infections, 157

Hepatitis, 64, 127, 161–7
  hepatitis A, (infectious), 8, 162–3
  hepatitis B, (serum), 35, 41, 42, 93,
    118, 126, 163–6, 193, 197
  spread, 2, 3, 9, 10, 11, 17, 55,
    119, 144, 163–4
  non-A, non-B, 166–7
Herpes simplex virus (HSV), 21, 28,
    103, 142, 161, 203, 205, 206,
    228
  genital, 178, 180, 181, 182, 183,
    188–9
  spread, 9, 18, 142, 143, 144, 191,
    195, 197
  therapy, 93, 94, 95, 130, 143
Herpesvirus simiae, 209
Herpes zoster see Varicella zoster
Herpetic
  keratitis, 236, 237
  stomatitis, 142–3
  whitlow, 143
Histoplasma spp, 3
Human immunodeficiency virus (HIV)
  (see HTLV-III)
HTLVs (Human T-lymphotrophic
    viruses), 128–30
HTLV-III, 11, 55, 93, 95, 127,
    128–30, 144, 181, 209
Human papilloma virus (HPV), 182,
    189, 232–3
Humoral antibody response, 33–4
Hyaluronidase, 22
Hypochlorites, 48, 52, 55

Idoxuridine (IDU), 94
Immune response, 24–5, 26–7, 31–8
Immunisation, 12, 36, 38–42
Immunity
  active
    acquired, 36–42
    cell-mediated 34–5, 36
    humoral, 33–4
    innate, 29
  passive
Immunoassays, 68, 69–70, 72, 112,
    147
Immunofluorescence tests
  bacterial, 69, 121, 184, 186
  chlamydial, 112
  viral, 68, 71, 104, 109, 128, 143,
    208
Immunoglobulins, 37–8
  levels after burning, 225
Immunopathology, 35–6
Impetigo, 9, 174, 227
Incubation period, 2

Infantile
  gastroenteritis, 152–3, 196
  paralysis, 205
Infections
  burns, 225–6
  cardiovascular system, 118–31
  central nervous system, 199–215
  control, 11–12
  diagnosis, 57–73
    protocols of investigation,
      62–6
  endemic, 4–5
  endogenous, 1–2
  exogenous, 2
  exotic, 5
  gastrointestinal tract, 145–67
  generalised, 15–17, 18
  genital tract, 177–89
  hospital cross infection, 225, 226
  localised, 15–16, 17–18
  locomotor system, 216–20
  mouth, 132–44
  opportunistic
  persistent, 27–8
  prevention, 29–56
  respiratory tract, 5–7, 97–117
  skin, 221–34
  sources and spread, 1–12, 15–18
  susceptibility, 12, 13, 24–5
  transmissibility, 14–15
  treatment, 74–96
  urinary tract, 168–76
  wound, 225, 226
Infectious hepatitis (hepatitis A), 8,
    162–3
Infective endocarditis, 63–4, 120–2
Infective mononucleosis, 126, 127–8,
    161
Inflammatory response, 21, 26, 30–1
Influenza viruses, 5, 24, 98, 99, 103,
    104, 105
  antigenic drift and drift, 108
  clinical syndrome, 107–9
  control, 41–2, 108–9
Injection of infectious agents, 9–10
Innate immunity, 29
Insects as disease vectors, 8, 10
Interferons, 21, 31–2, 95–6
Intranuclear inclusions, 21
Intra-uterine device (IUD), 177, 180
Iodine, 48, 52, 54, 55
Iodophors, 52

Kahn test, 185
Kanamycin, 88
Kaposi's sarcoma, 129, 130

Kelsey-Sykes test, 50
Keratitis, 65, 236–7
Keratoconjunctivitis, 235
Klebsiella spp, 22, 88, 110, 118, 145, 168, 172, 191, 195
Koch-Henle postulates, 66
Koch-Weeks bacillus, 235
Koplik's spots, 100
Kuru (see central nervous system)

Laboratory control of antibiotic therapy, 80–1
Laboratory diagnosis, 57–73
  interpretation, 72–3
  methods, 67–72
  protocols of investigation, 62–6
  specimens, 57–62
Lachrymal gland infection, 234
Lactobacilli, 135, 136, 137, 145, 168, 177–8
Laryngo-tracheo-bronchitis, 104–5
Lassa fever virus, 5, 127
LAV see HTLV-III
LDso, (median lethal dose), 13
Lecithinase, 23
Legionellae spp, 65, 90, 92, 112
Legionnaires' disease, 112–13
Leprosy, 240
Leptospiral infections, 65, 241–2
Leptotrichia spp, 134, 136, 137, 141
Leukaemia, 128
L-forms, 172
Lincomycins, 91
Listeria monocytogenes, 31, 130, 191, 201, 242
Liver
  abscess, 159
  infections, 161–7 see also Hepatitis
Lobar pneumonia, 109, 110
Localisation of viruses, 17–18
Localised infection, 15–16, 17–18
Locomotor system infections, 216–20
  defences, 216
  potential pathogens, 217
Louping ill virus, 204
Louse-borne fever, 244, 245
Lung abscesses, 110, 116
Lymphadenopathy, 127, 129
Lymphatics, spread of infection, 18
Lymphocytic choriomeningitis, 203
Lymphogranuloma venereum, 180, 181, 182, 186–7
Lymphomas, 127, 128, 130
Lymphoreticular system, and viruses, 126–30
Lysozyme, 30

Malaria, 5, 10, 17, 126
Malignant disease, 13, 25, 161, 232
Malignant pustule (anthrax), 239
Malnutrition, 25
Mantoux test, 115
Marburg virus, 127, 230
Measles virus, 100, 104, 110, 191, 207, 212, 214, 229–30
  vaccine, 39, 41, 230
Meningitis, 130, 196, 241, 242
  acute, 199, 200–3
  and CSF, 65, 200, 202, 204
  tuberculous, 202, 203
  viral, 188, 203–4
Meningococcaemia, 196
Meningococcal meningitis, 201–2, 203
Methicillin, 84
Methisazone, 95
Metronidazole, 92
Mezlocillin, 85
Microscopy, 67, 70
Microsporum spp, 3
MLD (minimum lethal dose), 13
Molluscum contagiosum virus, 60, 182, 228, 233–4
Monkey pox virus, 229
Mouth infections, 132–44
  common oral infections, 140–5
  defences, 134–5
  dental caries, 135–7
  normal flora, 132–5
  periodontal disease, 137, 139–40
  potential pathogens, 138
MSSU, 170, 171, 173
Mucosae, infection through, 9, 15–16, 19, 24
Mumps virus, 143–4, 175, 181, 191, 203, 234
  vaccine, 41
Murine typhus, 246
Mutation to drug resistance, 78–9
Myalgia, epidemic, 219
Mycobacterium spp, 35
  leprae, 240
  tuberculosis, 7, 16, 17, 18, 31, 113–16, 125, 130, 173, 181
Mycoplasma spp, 89, 90, 99, 133, 168, 172, 191
  pneumoniae, 103, 111, 126
Myositis, viral, 219

Nalidixic acid, 92
Necrobacillosis, 102
Negri bodies, 21
Neisseria spp, 83, 118, 125, 133, 136, 139

gonorrhoeae, 103, 181, 182, 183–4, 195
meningitidis, 16
Neomycin, 88
Neonatal infections, 195–8
  bacterial, 133, 183, 195–6, 235–6
  chlamydial, 112, 187, 196
  fungal, 140, 197
  prevention, 178, 197
  viral, 125, 129, 161, 166, 188, 189, 197, 236 see also Pregnancy
Nerves, spread of infection, 18
Netilmicin, 89
Neuraminidase, 23, 24
Nitrofurantoin, 92
Non-gonococcal urethritis (NGU), 182
Non-specific urethritis, 181
Novobiocin, 92
Nutrition, and infection, 25

Occupational hazards, 25, 55–6, 111, 143, 158, 163–4, 204, 208, 228, 240
Oncogenic viruses, 21
Opthalmia neonatorum, 235–6
Oral cavity see Mouth infections
Orbital cellulitis, 234
Orchitis, 181
Orf virus, 234
Osteomyelitis, 65, 218–19
Otitis, 99–100, 227–8

Pandemic infections, 5
Panophthalmitis, 237
Papilloma viruses (HPV), 182, 189, 232–3
Papovaviruses, 130, 212
Parainfluenza viruses, 98, 99, 104, 109, 197
Paralytic disease, 204–5
  general paralysis of the insane (GPI), 185
  infantile paralysis, 205
Paramyxoviruses, 229–30 see also Measles
Paranasal sinuses, 99
Paratyphoid, 40, 150–2
Paronychia, 227, 228
Parvoviruses, 126, 229
Pasteurella spp, 244
Pasteurisation, 8, 43
Pathogenicity, microbial, 13–28
Paul-Bunnell test, 128
Pelvic
  abscess, 160
  inflammatory disease (cellulitis), 180–1, 187
Penicillins, 83–6
Periapical abscesses, 139, 140
Pericarditis, 125
Perinatal infections, 129, 163, 183, 195
Perinephric abscess, 175
Periodontal disease, 136, 137, 139–40, 142
Peritonitis, 160–1
  pelvic, 180–1
Persistent infection, 27–8
Persistent generalised lymphadenopathy (PGL) (see AIDS)
Phagocytosis, 30–1
Phenolic compounds, 50, 52–3
Phenoxymethyl penicillin, 84
Phosphonoformate, 95
Picloxidine, 51
Piperacillin, 85
Pityrosporum, 226
Plague, 15, 242–3
Plasmodium spp, 126
Platelets, and viral infections, 127
Pleurodynia, 219
Pneumococcal meningitis, 202
Pneumocystis carinii, 128, 130
Pneumonia, 65, 109–11, 112, 130
Pneumonic plague, 242, 243
Polioviruses, 2, 17–18, 21, 203–4, 204–5
  vaccines, 39, 40–1, 205
Polyoma virus, 214
Polypeptides, 91
Poxviruses, 14, 182, 233–4
Pregnancy and infections, 11, 37, 41, 163, 189, 191, 192–6, 205
  vaccines, 38, 41, see also Congenital infections; Neonatal infections
Prevention of infections, 29–56
  disinfection, 47–56
  immune response, 24–5, 26–7, 31–8
  immunisation, 12, 36, 38–42
  inflammatory response, 21, 26, 30–1
  sterilisation, 42–7
Progressive multifocal leukoencephalopathy, (PMLE), 214
Prophylactic use of drugs, 76–7, 93
Propionibacterium, 137
Prostatitis, 181, 187
Proteus spp, 20, 118, 145, 168, 172, 227
  and drugs, 84, 85, 86, 88, 92
Protozoa, 11, 133, 136, 157, 182, 223

254 INDEX

Pseudomembranous colitis (PMC), 157
Pseudomonas spp, 24, 118, 145, 172, 191
  and drugs, 83, 85, 86, 87, 88, 90, 91
  aeruginosa, 91, 110, 168, 195, 223, 225, 226, 227, 235
Psittacosis, 112
Puerperal infections, 191-2
Pulp infections, 139-40
Purine nucleosides, 93-5
Purpura, 126
Pyoderma, 174-5
Pyogenic liver abscess, 159
Pyrexia, 62, 191
Pyrexia of unknown origin, 245
Pyrimidine nucleosides, 93-5
Pyuria, 172, 173

Q fever, 71, 77, 111, 120, 245
Quaternary ammonium compounds, 53, 54

Rabies virus, 18, 21, 93, 205, 207-9
  vaccine, 41, 208
Radiation for sterilisation, 46
Radioimmunoassay (RIA), 68, 69, 72
Ramsay-Hunt syndrome, 232
Rashes, 35, 65, 226-34
Rat-bite fever, 243-4
Recreational drugs, 10, 120, 129, 163
Reiter protein CF test (RPCF), 186
Relapsing fever, 10, 244-5
Renal infection, 170, 173-5, 241
Resistance to drugs, 78-80
  see also Specific drugs
Respiratory syncytial virus (RSVirus), 98, 99, 104, 109, 197
Respiratory tract infections, 97-117
  defences, 97
  normal flora, 97
  paranasal sinuses, 99
  potential pathogens, 98
  spread, 5-7, 11
  viral, 17, 93, 97-9
Retroviruses (see HTLVs)
Rhabdovirus *see* Rabies virus
Rheumatic fever, 77, 120, 123-5
Rhinoviruses, 98, 99, 103, 104
Rice water stools, 156
Rickettsial infections, 10, 90, 228, 245-6
Rideal-Walker test, 50
Rifampicin, 91-2
Ringworm, 9
Rotaviruses, 147, 197

Rubella, 127, 161, 207, 212
  and pregnancy, 11, 191, 193-4, 214, 236
  vaccine, 39, 41, 194, 218

Sabin vaccine, 40-1
Salivary glands, 143-4, 207
Salk vaccine, 41
Salmonella spp, 8, 18, 84, 88, 118, 130, 196, 201, 218
  paratyphi, 150
  typhi, 29, 150
  typhimurium, 29, 148
Salmonellosis, 148, 150
Salpingitis, 180, 187
Saprophyticus spp, 168, 172
Sarcinae spp, 221
Scabies, 182, 183
Scarlet fever, 227
Schick test, 102
Scrapie, 215
Scrub typhus, 245, 246
Secretions, bacteriocidal, 29-30
Septic
  abortion, 191
  arthritis, 216-18
  shock, 20
Septicaemia, 17, 20, 65
Serological tests, 68, 69, 71-2
  specimens, 60
Serum hepatitis *see* Hepatitis B
Sexually transmitted diseases, 169, 180, 181-9 *see also* AIDS
Shigella spp, 15, 19, 20, 84, 154-5
Shingles, 231-2, 236
Shipyard eye, 235
Sickle cell disease, 126
Sissomycin, 89
Skin infections, 17, 221-34
  defences, 222
  disinfection, 54-5
  normal flora, 221-2
  potential pathogens, 222-3
  specimens, 60
  spread of infections, 9-10, 15-16
  viral infections, 228-34
Slow virus infections (*see* central nervous system)
Smallpox, 12, 17, 95, 229
Sore throat, 7, 100-4
  investigation, 66
  sore throat syndrome, 102-4
Sources of infections, 1-12
Specimens, collection, 57-62
Spectinomycin, 89
Spirillum minus, 243, 244

Spotted fevers, 228, 245-6
Spread of infections, 1-12, 15-18
Sputum specimens, 61
Staphylococci spp, 22-3, 117, 120, 139, 168, 177, 178, 180
  albus (epidermidis), 118, 121, 133, 168, 172
  aureus, 22, 118, 120, 125, 201, 202, 209, 216, 219, 235-6
  and drugs, 83, 84, 88, 90, 91
  gastrointestinal tract, 8, 19, 146, 148, 149, 159
  genito-urinary tract, 139, 177, 178, 180, 192, 195
  respiratory tract, 99, 107, 116, 117
  skin, 221, 222, 223, 225, 226-7
Staphylokinase, 23
Sterilisation, 42-7
  filtration, 46-7
  gas, 47
  heat, 42-6
  radiation, 46
Streptobacillus moniliformis, 243, 244
Streptococcal sore throat, 15, 102-4, 123-5, 174-5
Streptococcus
  beta-haemolytic, 90, 99, 102, 104, 177, 192, 209, 216, 218, 223, 225-6
  faecalis, 118, 120, 145, 148, 159, 160, 168, 172, 177, 192
  and drugs, 75, 84, 87, 88, 92
  milleri, 159, 209
  mitis, 1, 120, 121, 134
  mutans, 24, 133, 134, 135, 136, 137
  pneumoniae, 105, 107, 109, 110, 116, 118, 201, 209, 218, 227
  pyogenes, 7, 8, 90, 102, 103, 123-5, 174, 201, 202, 227
  pathogenicity, 19, 22, 23, 24, 77, 103, 116
  salivarius, 133
  sanguis, 1, 120, 133, 134, 135, 136
  viridans, 1, 120, 136
Streptodornase, 23
Streptokinase, 23
Streptomycin, 87
Stye, 227
Subacute sclerosing panencephalitis (SSPE), 214, 230
Subphrenic abscess, 159-60
Sulphonamides, 81-2
Sulphur granules, 238
Surgical specimens, 61, 67
Swabs, diagnostic, 58-9, 60, 62
Symptoms: production, 27

Synergism, 75, 83, 87
Syphilis, 11, 66, 182, 183, 184-6, 193, 218, 228

TAB vaccine, 152
Talampicillin, 85
Tetanus, 15, 19, 36
  immunisation, 39, 211-2, 213
Tetracyclines, 89-90
Thrush, 130, 140, 182, 183, 197
Ticarcillin, 86
Tick-borne fever, 244, 245
Tine test, 115
Tobramycin, 88-9
Togaviruses, 10, 17, 119, 126, 203, 204, 206-7, 218, 219
Toxic epidermal necrolysis, 227
Toxic shock syndrome, 178
Toxigenicity, 18-21
Toxoid, 20, 36, 38, 102
Toxoplasma gondii, 11, 130, 193, 236
Toxoplasmosis, 127, 193, 209
Transmissibility of infection, 14-15
Trench fever, 10, 245-6
Trench mouth, 140-1
Treponema
  pallidum, 66, 182, 184-6
Trichomonas spp, 133, 178, 180, 182, 183
Trichomoniasis, 92, 182
Trichomycosis axillaris, 228
Trimethoprim, 82-3
Triple vaccine, 39, 40
Trypanosomiasis, 10
Tuberculin test, 114-15, 230
Tuberculosis, 75, 87, 113-16, 172, 173-5, 218
Tuberculous meningitis, 115, 202, 203
Tularaemia, 245
Tumour production, 127, 130
Tyndallisation, 43-4
Typhoid fever, 3, 8, 14, 40, 150
Typhus fever, 10, 245-6

Unconventional agents (see Kuru and Creutyfeld-Jakob disease)
Ureaplasmas, 168
Urethritis, 172, 181, 187, 188
Urinary tract infection, 168-76
  defences, 168
  drugs, 77, 82, 86, 88, 92
  normal flora, 168
  potential pathogens, 169
  viral infections, 175-6
Urine specimens, 61, 170-1, 173

INDEX

Vaccines, 38–42
Vagina, normal flora, 177–8
Vaginitis, 178
Vancomycin, 92
Varicella zoster, 18, 28, 33, 94, 95, 130, 197, 207, 231–2, 236
  immunisation, 39, 41, 232
Variola *see* Smallpox
VDRL tests, 185
Veillonella spp, 133, 136, 139, 209
Vibrio spp, 137, 141
  cholerae, 19, 155–7
  parahaemolyticus, 148, 150, 155–6
Vidarabine, 94
Vincent's infection, 140–1
Viraemia, 17
Viral pathogenicity, 21, 26–7, 28
  adherence, 24
  localisation, 17–18
Virulence, 13–14
Vulvitis, 178
Vulvo-vaginitis, 178, 183

Warts, 9, 15, 228, 232–3
  genital, 182, 183, 189, 233
Wasserman reaction, 72, 185
Water contamination, 8, 10, 112, 155, 156, 157, 162
Weil's disease, 241
Whooping cough, 6, 39, 40, 49, 60, 106–7
Woolsorter's disease, 239
Wound contamination, 9, 208, 211–12, 213
  infections, 223–5

Yeasts, 133, 177, 178, 180
  see also Candidiasis
Yellow fever, 10, 41, 162
Yersinia
  enterocolitica, 157
  pestis 18, 242–3

Zoonoses, 3, 158, 239, 241
Zoster (shingles), 231–2, 236
  see also Varicella zoster